ANCIENT SUNSHINE

The Story of Coal

*Coal miner's child carrying Kerosene
for lamps at Pursglove camp
Scott's Run, West Virginia*

ANCIENT SUNSHINE
The Story of Coal

By

James B. Goode

The Jesse Stuart Foundation
Ashland, Kentucky
1997

Dedication

To Donna Slone, my bride,
who patiently listens, offers superb advice,
proffers enthusiasm and always revels in my successes—
as I do hers. . .

ANCIENT SUNSHINE
The Story of Coal

Library of Congress Cataloging-in-Publication Data

Goode, James B., 1948-
    Ancient Sunshine : the story of coal / by James B. Goode.
       p.   cm.
    Includes bibliographical references.
    ISBN 0-945084-64-1
    1. Coal mines and mining.   I. Title.
TN802.G53 1997
622' .33--dc21

97-19063
CIP

Published by:
The Jesse Stuart Foundation
P.O. Box 391
Ashland, KY 41114
(606) 329-5232

Foreword

More than sixty years ago, the song "Which Side Are You On?" emerged from a National Miners Union meeting to become a symbol of the traditional conflict between coal miners and coal operators throughout Appalachia. That same conflict is also found in accounts of coal mining. Most people who write about coal mining are telling you in effect which side they are on. Books about coal mining are, therefore, often more about politics than they are about processes.

A former student Carolyn Traum called me three years ago and told me how much school children needed a book focused more on coal and coal mining. "We need a book," she said, "that tells young readers what coal is, and how it is mined, and how it is used." Carolyn is an educational leader who knows the needs of students in Eastern Kentucky and Southern Appalachia, so I took her advice seriously and proposed the idea to my friend James B. Goode.

Goode, a recognized authority on coal mining, was born and grew up in Benham, Kentucky, a coal town built by International Harvester. After I outlined the type of book we needed, he readily agreed to write the text and gather the photographs. He spent parts of the next two years preparing and refining a book that provides basic facts about Appalachian coal mining.

When James Goode completed his manuscript, our chief illustrator and book designer,

Malcolm J. Wilson, Photographer

Jim Marsh, began the complicated process of designing the book and preparing it for the printer. Jim had grown up in Twin Branch and Capels, coal camps in West Virginia, so his background gave him a personal feeling for the text and the photos. The book you hold in your hands is a happy combination of the writing skills of James Goode and the great

*Jim Marsh,
age 7*

creative abilities of Jim Marsh.

Coal mining in Appalachia began almost two hundred years ago and has changed dramatically since that time. Before the Civil War, coal from Appalachia was shipped by river as were most other goods. After the war, the growth of new factories and the spread of railroads greatly increased the demand for coal as a fuel. Railroads went into the coal fields to bring out the coal, and machinery made it easier to produce the large quantities required by an expanding nation.

During the twentieth century, coal mining experienced further changes. Unions developed and later declined. Mechanization changed the way that coal was mined. New laws placed greater emphasis on miners' safety and the reclamation of mined land. With mechanization and declining markets, mining families moved north in search of steady work.

Rapidly changing coal-mining technology signals dramatic change in the industry's future. But understanding the past helps us build bridges from the present to the future.

James M. Gifford
March 20, 1997

Special Thanks to:

~David & Susie Duff, Pine Branch Coal Sales

~Becky Farley, Technical Information Specialist

~Learning Resource Center National Mine Health and Safety/ Beckley, West Virginia

~The University of Kentucky/ Southeast Community College/ Appalachian Archives

~The University of Kentucky/ Margaret I. King Library/ Special Collections

~The Arch Mineral Corporation/ Number 37 Mines/ Cumberland, Kentucky

~Dean Cadle, Photographer

~James Larkin Goode, Photographer

~Malcolm J. Wilson, Photographer

~Betty Jean Hall, Judge

~Libby Lindsay, West Virginia Women Miner's Support Group

~James G. Clough, Alaska Division of Geological & Geophysical Surveys

~David Todd, Arch Coal, Inc.

~Carolyn Kersey, Ashland Inc.

Contents

*Pupils of the Lynch Graded School
gather for lighting of Christmas tree
Lynch, Kentucky, 1920*

Introduction

The use of fire for heat, light, and invention is one acquired skill that has made man distinct from all the other animals. Without fire, and his use of it to forge simple metals, man would not have risen to his sophisticated level of technology. Think of all the components in a computer that are made from forged or cast metals. The use of coal to produce a hotter, more intense fire with which to forge harder metals such as cast iron and steel has become the foundation for man's unique technological achievements.

In the United States, coal is our most abundant fossil fuel. In recent years, we have developed more dependence upon oil and nuclear energy. Coal has also gained in importance as the world has tried to find a way to solve its increasing energy problems. Man's inventiveness and experimentation has also produced hundreds of by-products which has increased the importance of coal in a modern society. Although we know a great deal about coal, there are still lots of things we have yet to discover. We can learn to use it more effectively and in an increasing number of ways.

My first contact with coal was as a young boy growing up in a coal camp in Eastern Kentucky. Both my grandfathers worked for International Harvester Company/ Wisconsin Steel Corporation at Benham, Kentucky. My Dad began working for the Harvester Company in the 1930s and retired in 1972, after 40 years of service.

We burned coal in our grates and stoves in order to heat the house and to cook. One fancy coal heating stove sat on a stove board in the living room. The kitchen contained a smaller laundry stove and a large, white, porcelain cook stove. The laundry stove was equipped with a hot water jacket which was a set of hollow, rectangular, cast-iron containers serving as reservoirs for water which circulated through, was heated, and then stored in a tall, galvanized water tank behind the stove. We used this hot water for everything from bathing to scalding chickens.

The stove in the living room was our primary source of heat for the house. We had our coal delivered by the company and shoveled into a spacious, oak coal bin which sat along the alley behind our house. Back then, one could order either "block" or "egg" coal. Block coal was cut into large squares which could measure up to one foot on a side and required the use of a coal ax to chop the coal into manageable pieces. My Daddy preferred egg coal which was smaller and varied from golf ball sized lumps to some the size of a medium cantaloupe.

We carried the coal to the house in five gallon lubricant buckets which had been discarded by the coal company. When the coal was first delivered, the surfaces glistened in the sun. I always thought that the lumps looked like onyx jewels. After a while, the coal oxidized and the surfaces turned dull. As we used the coal ax, the coal would be split open to reveal the same, bright, glistening surface. I carried four buckets at a time—an effort my Daddy characterized as "a lazy man's load." Once every morning and night I carried the coal buckets to the house.

4th Of July Parade By Employees of
U. S. Coal & Coke Company
Lynch, Kentucky, July 4, 1925.

We also used coal to feed hogs. My Daddy said that the sows needed "tankage" and would eat their little ones if they didn't have carbon. We broke the coal into small pieces and the hogs crunched the coal like ice cubes.

Some of my neighbors, who had forges, burned the coal to heat metals to make them malleable. The metals had to be heated to make repairs or create new devices such as specialized tools. Their forges contained hand cranked blowers which allowed air to be forced across the burning coal creating a hot fire which brought the metals to the correct temperature.

Coal fires created lots of smoke in the camp. Imagine over 600 households burning coal in their grates, heaters, and cook stoves! The soot from the chimneys rolled from the tar-paper-covered roofs like black snow. I found out later that we had a form of acid rain which hung over the camp. When the sulfur dioxide created by the coal smoke met the moisture in the atmosphere, sulfuric acid was formed. I remember my mother's fiberglass draperies disintegrating from the acid in the air. The acrid, yellow, heavy smoke stung my nostrils. I shall never forget the sulfur smell in the house created when high winds would force our old coal stove to puff smoke into the room.

The coal left our camp by rail on gondolas pulled by giant 500 ton steam engines. I recall the long strings of cars and the vibrant, thunderous roar of the powerful engines as they left town in a cloud of billowing, drifting smoke.

Our camp supplied all the coal used by Wisconsin Steel Corporation in making steel for International Harvester's products. Trucks, tractors, bulldozers, refrigerators, and hundreds of other products were made from the raw steel.

WHAT IS COAL?

Coal is one of the most complex substances on Earth. Scientists classify it as a solid, brittle, combustible rock that is made primarily of carbon. It does contain other substances classified into silicates, oxides, carbonates, sulfides, sulfates, phosphates, and arsenides. These groups contain chemicals such as alkalis, calcium, magnesium, iron, titanium, and silica.

HOW IS COAL FORMED?

Understanding how coal is formed will help us consider the complexity of this ore. Many of us grow up believing that coal simply was formed by plants falling into muddy swamps and being covered by silt that eventually created pressure and heat and pressed the carbon into seams. Some of this is true but the process is much more complicated.

Coal formed from plant substances. These plants stored the sun's energy during the process of photosynthesis. They were preserved from total decay by being placed in a favorable environment where various chemical and physical processes converted this stored energy into coal. Scientists estimate that photosynthesis in the oceans produces over 12 million tons of hydrocarbon material annually. Just a small fraction of this would produce all of the fossil fuel we currently know about in the world. Scientists have calculated that it takes 20 feet of vegetal matter to form 1 foot of coal.

Coal generally formed near the ocean shore in an environment of low topographic relief that was exceptionally favorable for plant growth. These were places where trapped water and sediment from frequent flooding could accumulate. These places experienced very stable environmental conditions over a long period of time. There was a tropical climate, an atmosphere of high humidity, and the presence of dissolved materials typically found in ocean water.

Many of the plants which make up the content of coal seams may be found as fossils in the overburden. Some of these giant ferns, such as sigillaria, reached heights of 150 feet. Scale trees called lycopods, seedferns called pteridosperms, horsetail rushes called sphenopsids, and cordiates,

James B. Goode, Photograph

Coal is stockpiled, then loaded into rail cars at
Pine Branch Coal Company Hoyt Tipple
Chavies, Kentucky, April 29, 1995

which are ancestral to conifers, are common. It is not unusual to find outlines of sea life in the gray slate. Formanifera, corals, mollusks, brachiopods, bryozoans, crinoids, and trilobites are occasionally seen. Although the torrid heat and carbon-dioxide-charged atmosphere would not support air breathing organisms during this time, amphibians, fishes, clams and other shellfish, and giant dragonflies with 30 inch wingspans were in abundance.

All this starting material for coal forms peat which is a combustible, brown, loosely compacted material often cut into blocks to burn in fireplaces—especially in Europe and the British Isles. Under favorable conditions, one foot of peat may form every ten years. As physical and chemical processes begins to act upon the peat the material turns to lignite, to bituminous coal, and then to anthracite. This formation began to occur during the Carboniferous Period over 300 million years ago during a very slow process.

Most of the high-grade coal deposits in the eastern United States were formed during the Mississippian and Pennsylvanian ages, 270 million to 350 million years ago.

Coal is found in very distinct strata or seams between layers of inorganic rocks. These seams range from a few inches to several feet thick over areas that extend hundreds of square miles. Sometimes there are as many as thirty seams stacked like layers of a cake. Some areas only have one seam. For example, the deposits found throughout western Montana and Wyoming are in one seam which sometimes measures 300 feet thick. The largest coal seam is in China and is over 400 feet thick! The average coal seam in the United States is 5.4 feet.

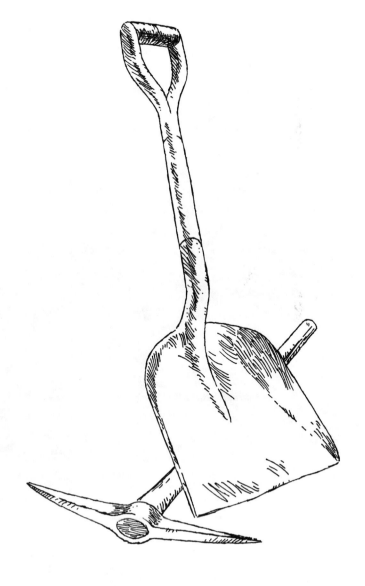

A Miner's basic tools - the pick and shovel

This coal powered steam shovel
cleans debris from a coal tipple at
Lynch, Kentucky, 192?

Austin Morgan in machine shop
Benham, Kentucky, October 8, 1904

Varieties of Coal

The depth of burial and the time of compaction generally determines the type of coal.

There are four types of coal:

Lignite

- Is the youngest and least "mature" on the coal scale

- Contains visible traces of original vegetable matter.

- Ranges in color from brown to black and contains 30% fixed carbon

- Has a low heat value but burns with very little smoke

- Contains too much moisture—sometimes up to 50%!

- Is subject to rapid decomposition when exposed to oxygen

- Is subject to spontaneous combustion

- Abundant, reserves are found in the West and Gulf Coast regions of the United States

- Largest deposits of lignite in the U. S. are in western North Dakota and eastern Montana

- Cannot be easily transported because of its combustibility

- Produces about 7,000 B.T.U.'s per pound

Sub-Bituminous

- Has a high-heat value greater than lignite

- Sometimes called "black lignite" but ranges in color from brown to black

- Has a substantially lower moisture content than lignite

- Is "woody" and gives off a great deal more smoke than lignite

- Reserves are found mainly in Wyoming, Washington, Utah, New Mexico,Montana, and Colorado

- Is mainly used for steam production

- Produces about 9,500 B.T.U.'s per pound

Many coal camps generated their own electric power. These men are firing furnaces for a steam powered electrical generator Benham, Kentucky, 193?

Bituminous

- The most plentiful of all coals

- Not all bituminous coal makes good coke

- All metallurgical coke used to make steel is made from bituminous coal.

- 54 percent of electricity in the U.S. is generated using bituminous coal.

- Ranges from dark gray to black in color

- Usually contains about 5% moisture

- Produces approximately 13, 000 B.T.U.'s. per pound

Anthracite

- Is more difficult to ignite but burns longer and cleaner

- Is easier to store because it is more resistant to deterioration

- Most anthracite deposits occur in western and northeastern Pennsylvania

- Has a carbon content of approximately 86%

- Burns slowly with a non-luminous flame that gives off almost no smoke

- Produces approximately 13,000 B.T.U.'s per pound

- Anthracite coal is the closest to pure carbon of any of the types of coal. If the proper conditions existed, more heat and weight for more time, it could turn to the pure carbon which is diamond. This has only occurred in a very few places in the world.

- Began to be used in forges as early as 1769 near Wilkes-Barre, Pennsylvania

- Records indicate that it was used at Carlisle, Pennsylvania to forge guns for the Continental Army during the Revolutionary War.

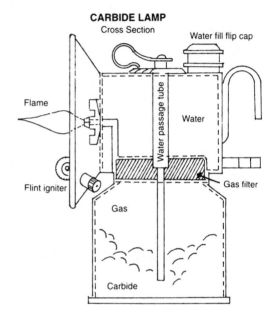

CARBIDE LAMP
Cross Section

Water fill flip cap

Flame

Water passage tube

Water

Flint igniter

Gas filter

Gas

Carbide

The Carbide Lamp

General Electric coal fired electrical generator
Benham, Kentucky, 193?

Yard engines such as this steam locomotive on track near
Benham, Kentucky were used to pull coal gondolas in the
rail yard.

Coal burned with limited oxygen produces coke for steel production. These coke ovens at Benham, Kentucky were active in the 1930s.

WHAT MAKES COAL BURN?

What is fire? This is one of the questions firemen are often asked when testifying at arson cases in courts-of-law. This definition is important in answering the question: "What makes coal burn?" We already know that coal is actually "ancient sunshine." Energy from the sun, absorbed by photosynthesis, is stored in the plant. As coal is formed, this stored energy is converted to carbon and other chemicals. Fire is actually a chemical reaction which is begun by the introduction of heat or friction. When oxygen or other combustible gases are combined with a liquid or solid combustible material such as coal and the right amount of heat or friction is introduced, a rapid loss of this stored energy occurs. Carbon is a material that oxidizes rapidly, once the process is initiated. The chemical reaction called fire occurs swiftly and releases both heat and light.

Even rust is a form of fire. Rust is a slower chemical reaction which is caused by combining oxygen and another combustible such as steel or iron. The rust causes the metal to slowly burn away.

USES OF COAL

Seventy-five percent of all coal mined in the United States is used for two purposes: generation of electricity and production of steel. Some coal is still used for heating homes and businesses. But some of the by-products of burning coal can produce a variety of plastics, nylon, synthetic rubber, paints, varnishes, drugs, and explosives.

Coal is the only known substance on Earth which can be made into coke. Coke is made by baking coal in a sealed oven and driving off the impurities. Coke is valuable because it makes a fire hot enough to turn iron ore into steel.

Generally, when coal is carbonized through the coking process it can produce five branches of products: 1) Gases; 2) Coke; 3) Tar; 4) Chemicals; and 5) Light Oils. Some of the most recognizable products are perfume, pipe stems, photo developer, baking powder, insecticide, moth balls, TNT explosive, graphite, paving, roofing, fertilizer, rubber stamp ink, lighter fluid, laughing gas, linoleum, laxative, disinfectant, wood preservative, and varnish.

The most frequent use of coal is by electric utility companies as a fuel to heat water, create steam, and turn turbines that generate the majority of our electricity. These utility companies consume 2/3 of all coal produced in the United States.

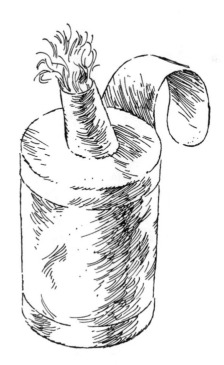

One of many types of mining lamps, this one uses oil and has an open wick.

19

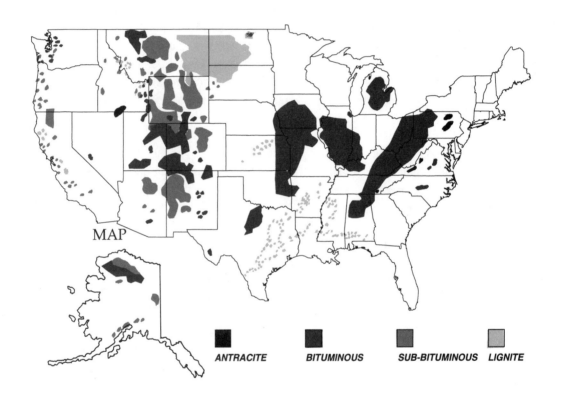

MAP

ANTRACITE   BITUMINOUS   SUB-BITUMINOUS   LIGNITE

Geography of Coal

The location of coal seams in the United States is divided into six major regions:

The Eastern Province

- Includes from Alabama, up the Appalachian Mountain chain, and ending in Pennsylvania.

- This is the most important coal mining region in the United States.

- Nearly all of the coking coal for steel making comes from this region.

- This area includes the largest deposit of Anthracite coal in the United States.

- Strippable reserves are estimated at five billion tons—two billion of this amount is classified as low-sulfur.

- Appalachian Region has the most highly developed coal infrastructure (developed coal mines, tipples, rail transportation, steel mills, and trained human resources) of all coal fields in the United States. A large quantity of these reserves is being held for future development.

The Gulf Province

- Includes Texas, Arkansas, Mississippi, Alabama, and Louisiana.

- Almost all these reserves are lignite.

The Interior Province

- Includes Illinois, Indiana, Michigan, western Kentucky, and coal deposits running from Iowa to Texas and Arkansas.

- This area is blessed with a variety of coals including sub-anthracites to high-volatile bituminous.

- Strippable reserves are estimated at eighty and a half billion tons with only about a billion tons of that are classified as low-sulfur.

Rocky Mountain Province / Northern Great Plains Province

- This includes from the Canadian border, down the Rocky Mountain range through Arizona and New Mexico.

- This area includes vast reserves of sub-bituminous and bituminous coal

- 25 billion of the 27 billion in strippable reserves are classified as low-sulfur

Pacific Coast Province

- This area runs along the western coast from Canada to Mexico

- This is the smallest of the United States coal fields at only 160 million tons of strippable coal

Alaskan Province

- This area has vast unexplored reserves. Geologists estimate reserves at 5.7 trillion short tons of bituminous coal—which was more than the total continental United States' identified resources in 1972. Coal-bearing strata underlie more than 9% of Alaska's land area. A significant portion of this deposit has coking qualities and is considered metallurgical. Additionally, there are some anthracite deposits. The bituminous deposits are divided into six basins: 1) North Slope (2,500 billion short tons) 2) Cook Inlet (500 billion tons) 3) Alaska Penisula (3 billion tons) 4) The Gulf of Alaska (4 billion tons) 5) Upper and Lower Koyukuk (1 billion tons) 6) Kobuk (1 billion tons). Additionally, there are 2,564 billion tons of sub-bituminous coal.

Facts About U.S. Coal

- Coal in eastern United States was formed approximately 300 million years ago; in the west, it was formed much later.

- The largest coal producers in the world are: 1) The United States; 2) The Soviet Union; 3) China.

- Coal is the most abundant U.S. fossil fuel.

- One-fifth of all accessible and usable coal is held by the United States in the lower 48 states (about 259 billion short tons).

- 99.9% percent of the land surface in the United States has never been touched by mining.

- Coal is the most difficult fossil fuel to recover, transport, and handle.

- The American coal miner produces more coal-per-man than any other nation in the world.

- One-fifth of all coal is surface mineable.

- Three-fourths of all surface mineable coal in the United States lies west of the Mississippi River.

- More than one-half of the total coal reserves in the United States requires underground mining, lies west of the Mississippi, and is bituminous.

- Montana and Wyoming hold more than a third of the total U. S. underground resources.

- Geologists estimate that the total U. S. resources of coal in the lower 48 states are roughly 3.2 trillion tons.

- The United States has more recoverable coal reserves than any other nation on earth, except for China.

- Four-hundred billion tons of coal reserves have been carefully measured and assessed as being exploitable under local economic conditions and available technology.

- Coal deposits in the United States are estimated to be far larger than all of the oil deposits in the Middle East.

- Coal is presently being mined in 27 of the 50 states. At least eight more states contain recoverable reserves.

- Currently, more than 300,000 people in the United States work in the mining industry.

- Coal production has risen from 610 million tons in 1969 to over one billion tons in 1994. Even at this rate, experts estimate that it would take 1,000 years to deplete these vast reserves.

- Coal is a $55 billion industry which generates 57 percent of electrical power in the United States.

- Japan is by far the largest customer in the Far East for U. S. coal and is dependent upon imported coal for both steam generation and metallurgical purposes.

- Canada is the largest volume purchaser of U. S. unprocessed coal.

"Coal Miner Smoking Pipe"
Williamson, West Virginia, 1935

This wooden car of coal was the first
shipped from U.S. Steel's mines at Lynch,
Kentucky to assist with the war effort
"First shot at the Kaiser from Lynch,
Kentucky, November 1, 1917"

Facts About Appalachian Coal

- Appalachian coal has been mined for over 200 years.

- More than 1/2 of the total U. S. coal mined comes from the Appalachian coal basin.

- The coal reserve base in Appalachia is the source of most consumption in the U. S.

- Two-thirds of the total Appalachian coal reserves is located in West Virginia and Pennsylvania and is mostly mined by underground methods.

- Ohio and Eastern Kentucky contain the next largest coal reserves.

- Slightly more than one-fourth of the underground and one-eighth of the surface reserves in the United States are located in Appalachia.

- Appalachian coal is lower in sulfur content, higher in BTU, and lower in ash than competing coals.

- Some of the best coal is found in Eastern Kentucky and the least desirable is in Ohio, followed by West Virginia and Pennsylvania.

- The Appalachian Region has the most highly developed infrastructure of all the coal mining areas of the United States.

Water sprayers were used to suppress coal dust in underground mines. This one sits in a mine drift mouth at Benham, Kentucky 192?

Facts About Kentucky Coal

The Kentucky Coal Facts: Pocket Guide published by Kentucky Coal Marketing and Export Council and the Kentucky Coal Association is an excellent source of information. The guide includes data and information about production, employment, economy, coal markets, environment, coal resources, and electricity. Additionally, there are short sections on the history of coal, types of mining, changes and trends, and a list of references. Two WEB sites are of particular importance in coal education: http://www.coaleducation (provided by the Kentucky Coal Association through a grant by the Kentucky Coal Marketing & Exporting Council) & http://www.miningusa.com (sponsored by Mining Internet Services, Inc. of Lexington, Kentucky).

Here is a summary of just some of the valuable data this source has to offer:

- Kentucky produced 20.8% of the nation's coal in 1970 but just 15.6% in 1994.

- Wyoming ranked first in coal production in 1994 with 237.1 million tons compared to Kentucky's ranking of third with 161.6 million tons.

- In 1994, Kentucky had 248 surface mines which produced 66.2 million tons of coal compared to 425 underground mines which produced 95.4 million tons.

- In 1994, Western Kentucky only had 24 underground mines compared to Eastern Kentucky's 401.

- Kentucky's largest coal producer in 1994 was Pike County with 56,908,972 tons.

- Kentucky coal mines employed 47,190 men in 1979. That fell to 23,368 in 1994.

- An emergency medical technician is required at underground coal mines employing 12 or more employees, with an additional EMT per each additional 50 miners.

- A minimum of 16 hours of annual retraining is required to maintain the miner certification and allow the miner to continue to work at an underground mine.

- Wages paid by the coal industry in 1994 totaled $942,817,789.

- Coal companies paid $158,114,435 in coal severance tax in 1994 on a gross value of the coal estimated to be at $3,546,899,946.

- In 1994, 95% of Kentucky's electricity was generated from coal.

- The largest customer for Kentucky coal in 1994 was Tennessee Valley Authority.

- Kentucky's 1994 export of 7.2 million tons were 10% of total U.S. exports.

- Kentucky has over 2,700 miles of railroad lines, over which 102.5 million tons of Kentucky coal were transported in 1994.

- Airports, mountain top farms, structural building sites, correctional facilities, industrial/commercial uses, government facilities, and R&R sport uses top the list for postmine land uses in Kentucky.

- As of 1994, Kentucky Harlan and Pike counties led the state in remaining coal resources. Experts estimate that 54.1 billion tons of reserves remain in Eastern Kentucky and 36 billion tons in Western Kentucky.

History of Coal

Prehistoric humans most certainly saw coal but might not have known its usefulness.

Coal "outcropped" in many places along stream beds and man began to experiment with it as a resource. History records use of coal by the Welsh at Glamorganshire over 4,000 years ago. They used it to cremate the dead in their stone encased funeral pyres. Later, when the Romans occupied portions of the British Isles, they mined and used coal. Roman soldiers were particularly fond of camping near coal outcroppings and building fires with which to keep warm.

Marco Polo, the great Venetian explorer, discovered that the Chinese burned coal to heat bath water and to make metal malleable. Coal is mentioned several times in the Bible. American Indians used coal to fire their pottery furnaces.

Because of the abundance of wood, coal was first thought of as an alternative fuel. As wood resources began to dwindle, particularly in England, coal began to be used more and more. Since coal had played a role in heating metal for hundreds of years, it was natural that it be utilized more as the world's population began to grow and technology, which required metals, began to develop.

One of the most significant developments was the invention in the 1800s of the Bessemer furnace which was designed to heat iron ore and turn it into steel.

Many of our inventions have come as a result of the pursuit of mining and processing coal. Scientists are continually discovering new properties and uses of coal. Today, the by-products of coal number into the hundreds. Mining of coal, as well as other minerals, directly affects us all. Your television set contains 35 different minerals and elements; a telephone contains 40; and a computer requires 32. Many of these are contained in plastics, steel, or other materials produced from processing coal.

Black Miner Loading Coal In Underground Mines Lynch, Kentucky, 192?

Early Coal Mines

■ Early mining did not involve sinking tunnels which followed the seams for any distance. Coal was "quarried" much like limestone still is today. A pit was dug where the coal outcropped at the surface and it was expanded a few feet down and then horizontally.

■ The first actual underground mines were shaped like a long-neck bottle. The miner entered the seam where it outcropped at the surface, sunk a shaft down several feet and then hollowed out the bottom of the shaft, expanding wider and wider until the conditions became unsafe.

■ As early as the fourteenth century, some coal was mined in seams worked horizontally. Miners did not go very far underground because of the lack of inventions to effectively prevent cave-ins.

There was a need for a variety of skills in underground mining. These miners are constructing a crib and overcast in underground mine Benham, Kentucky, 194?

■ By the early 18th century, horizontal shafts were extended much farther than previously. Miners used picks and shovels to excavate the tunnels. Timbers and close fitting wooden staves helped curb cave-ins. Men, coal, and supplies were moved by hand-operated windlasses.

■ Flooding in coal mines was first solved by using buckets, rope, and windlasses to extract water. Later, man-powered treadmills, gins turned by animals, water wheels, and mechanical pumps were used.

Coal is transported underground in a variety of ways. Pictured here is a shaker conveyer which advanced the coal by "shaking" it forward and back Benham, Kentucky, 194?

27

The Perils of Mining

Many experts say that, next to fire fighting, coal mining is the most dangerous profession on Earth. Many elements contribute to this: electricity, explosions involving gas or dust, explosives or other agents used to shatter coal, roof falls, bottom heaving, haulage, machinery, mine fires, pressure bumps or bursts, and sliding material.

Miners are electrocuted from working with high voltage electricity. Some are killed or injured in explosions involving volatile gases, some from falling or sliding rocks, some from handling explosives, some from machinery accidents which range from being pinned between machinery and the rib to getting a hand or arm caught in a conveyor belt. Sometimes, mine fires occur which cause respiratory injuries or result in death.

Mine disasters are usually defined as accidents that claim five or more lives. The worst years for mine disasters were between 1901-1925 when 305 coal mines reported accidents involving five or more fatalities.

The deadliest year in United States coal mining history was 1907, when 3,242 deaths occurred. That year, America's worst mine explosion ever occurred at Monongah Mines at Monongah, West Virginia, when 362 persons lost their lives in an explosion. The next most serious coal mine disasters occurred in 1913 at Dawson, New Mexico, when 263 men lost their lives in an explosion; and in 1909 at Cherry, Illinois, when 259 men lost their lives in a mine fire.

The most disastrous mine accident in recent history occurred at the Scotia Mine of the Blue Diamond Coal Company at Ovenfork, Kentucky in 1976 when 26 persons lost their lives in two successive explosions.

Improved technology, preventive programs, regulations, and the formation of regulatory agencies have drastically reduced the number of mine disasters or accidents. All coal companies either hold classes in safety and proper use of equipment for new miners or contract with outside professional agencies to conduct the training. Federal regulations require yearly re-training. Many companies have a safety department and hire safety engineers to ensure lower accident rates.

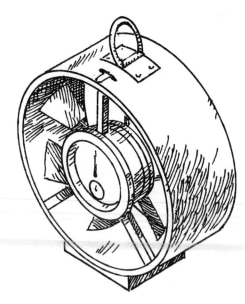

Anemometer - An instrument for checking and measuring the force or speed of the air in a mine

28

This mock accident scene at Benham, Kentucky demonstrates the hazards of working with coal cars. October 23, 1920

Some of the events that have really helped to assure the coal miner's safety include:

1910~ Congress creates the U. S. Bureau of Mines, charged with the responsibilities of investigating accidents, advising the industry, conducting production and safety research, and teaching courses in accident prevention, first aid, and mine rescue.

1969~ Congress passes the Federal Coal Mine and Safety Act which supports state and federal laws to better advise and regulate the mining industry, to extend coverage to all types of miners, to require or encourage use of successful safety procedures and technology, to provide effective miner training, and to focus on reducing or eliminating the most serious hazards.

1973~ Congress creates the Mining Enforcement and Safety Administration within the Department of the Interior with responsibility for safety and health enforcement responsibilities within the Bureau of Mines.

1977~ Congress establishes the Mine Safety and Health Administration within the Labor Department.

These coal miners take a break after their shift at Williamson, West Virginia, 1935

Coal companies are required to provide first aid training to their underground miners. Modern mines are required to employ Emergency Medical Technicians. Benham, Kentucky, 192?

Improvements in technology have greatly improved conditions in the mines. Major improvements in safety have included installing devices to detect and vent methane, adding protective cages to the drivers' compartments, deploying remote control operated machinery, and using longwall mining devices.

Mining methods have also improved. In underground mining, roof control has progressed from just timber support or wooden roofbolts to a combination of timber support and metal resin roofbolts or steel arches. Better drills, improved jacks, and less time spent under unprotected roof have helped.

There is more cooperation among management, unions, miners, and government. With all these entities working together, the number of injuries and fatalities has dropped sharply since that disastrous year of 1907.

This tired coal miner sits on the running board with his round, metal dinner pail nearby.

Farm Security Administration Project/ Marion Post Walcott

James B. Goode, Photographer

Benham Coal Company.
This bearded coal miner operates a piece of low seam mining equipment. 1983

Executive House
Dec. 11, 1918.

Coal companies often built fancy houses for their
corporate executives. This one was built by
International Harvester December 11, 1918 at
Benham, Kentucky.

Company Owned Coal Camps

Beginning in the 1880s, many business speculators developed partnerships with government officials and local leaders and managed to buy huge quantities of mineral rights cheaply. Later, they built railroads and opened up these territories to mining. There was a severe labor shortage in supplying these remote locations. Men had to be recruited to come to such isolated places.

Shortly after the turn-of-the-century, hundreds of coal camps were built along the narrow valleys. These camps varied in their size and quality. Many of the larger corporations built elaborate camps with state-of-the-art commercial and residential buildings.

Other small, family-owned operations could only afford a minimal investment.

For example, the houses at United States Steel Corporation's camp at Lynch, Kentucky, sat on native, cut-stone foundations, were constructed of kiln-dried lumber, and had Welsh-made slate roofs. In addition, a crude, but effective, sewer system carried waste away by using a stream of treated water which ran through every outdoor privy. At Inland Steel's Wheelwright, Kentucky, an insightful superintendent decided that coal smoke produced from heating stoves was unhealthy and proceeded to drill natural gas wells which were used to

Private Collection of James B. Goode

Coal Camps were often segregated along ethnic lines. Blacks who worked for The Black Mountain Corporation in Harlan County, Kentucky lived in this section of the camp. August 5, 1919

This was a typical cottage built by United States Steel at Lynch, Kentucky in the 1920s

heat the buildings. He also incinerated garbage and circulated heated water to all the residences.

By sharp contrast, there were coal camps on the west end of Harlan County, Kentucky, which had no running water in the residences and an overwhelming stench that encircled the camp from the festering outdoor privies. Coal smoke inundated the houses and made it almost impossible to clean house or wash clothes.

Many companies, in order to recruit and hold the men, built some conveniences. Hotels, schools, theaters, department stores, hospitals, clubhouses, bowling alleys, banks, and baseball fields were constructed. In some camps, miners were entertained, in person, by Gloria Swanson, Lash LaRue, Ken Maynard, or by a host of medicine shows, musicians, magicians, and opera stars.

Because most coal miners came to the camp out of poverty, they could not afford to purchase housing. Many who came were white families who left their failing mountain farms to find wage-paying jobs. The company built rental residences and the miners' pay was cut for the rent and electricity.

Miners paid for the number of "drops" in their house. These were electrical cords which came out of the ceiling in the middle of the room and the total number of these determined the bill.

The camps also differed sharply in the quality of other basic services including water, telephone, solid waste disposal, sanitation, medical care, and house maintenance. Some provided water from open, dug wells; others provided treated water in hydrants which served more than one house; some provided treated water piped directly into the house.

Most telephone systems were owned and operated by the company who maintained a switchboard and operator who completed calls. There were very few telephones in private residences. Sometimes community telephones were placed on the porch of a house located within a cluster of five or six residences.

The larger, corporate camps purchased a garbage truck and hired a crew to pick up the solid waste created by the camp households. Garbage was usually hauled to a large, un-regulated dump where it was covered with dirt. Smaller camps left the employees to their own devices for disposal. This usually

Many corporate coal camps had Vaudeville theaters. Seniors at the company High School perform a play in the one at Benham, Kentucky. 1925

meant that it was either buried in the backyard, thrown on the bank of a water way, or disposed of in one of several informally designated dumps along remote roads. Many of these dumps were thick with rodents. Some young men, who grew up in these camps, tell stories of taking their fishing poles to the company dump, baiting the hooks with cheese, and casting out into the piles of garbage. The rat would try to snatch the bait and, after the hook was set, the fight was similar to hooking a big fish. When they got bored, they would simply cut the line with a pocket knife.

Most all of these camps had outdoor toilets. The company contracted with a crew to remove the waste material and spray disinfectant to reduce the number of insects, rodents, and other vermin associated with unhealthy conditions. These crews, who often appeared in the shadowy dark late at night,

This coal miner takes a short break from his work at the D-7 underground mine at Benham, Kentucky. 1983

were called "honey dippers."

Some coal companies provided no hospital or medical personnel to care for their employees. Others built modern hospitals staffed with graduates of such prestigious institutions as Johns Hopkins University. Coal company doctors were paid a retainer and were allowed to collect some office call fees to supplement their incomes. This was, perhaps, the first managed health care system in the United States. These mountain physicians made door-to-door house calls to do everything from deliver babies to treat indigestion.

Since the houses were owned by the coal companies, they were responsible for maintaining the houses. Some built the homes of green lumber in a simple batten-strip style which caused them to have cracks after they seasoned. These homes often lacked insulation and were heated by one or two coal-burning stoves. In the smaller camps, many companies did only minor repairs to the homes. Most often these entailed roof repairs. Others built modern homes with lap siding and took care of the painting, electrical and plumbing repairs, and any carpentry which was required.

The quantity and quality of social and religious life in the camps varied. In several of the corporate camps, movie houses, theaters, parks, tennis courts, band stands, and athletic fields abounded. Certain ones had non-denominational church buildings where any religious organization could hold services by scheduling an appointment. The company often financially supported public celebrations of Independence Day, Thanksgiving, Christmas, and Easter. Some even celebrated the anniversary of the signing of the Emancipation Proclamation on August 8th. Theater productions, opera, brass band concerts, big band dances, and lectures were common.

Many family-owned camps depended upon the nearest merchant town to provide what little entertainment the miner's family enjoyed. This often involved events at commercial movie houses, drinking and gambling in the local taverns, or attending religious events in the many rural churches nearby. In one instance, a coal operator brought in a "girlie" show to distract miners from attending a union organizing meeting.

Certain coal companies built modern school systems for both blacks and whites in their camps. Although the school system

Kentucky Legion: Mooseheart Legion of the World at a Lynch, Kentucky, Fourth-of-July Celebration. 1927

was segregated, top quality teachers and administrators were hired to staff both schools. Students participated in a wide array of academic courses and extracurricular activities. Latin, home economics, industrial shop, band, dramatics, forensics, advanced mathematics, physics, and a host of other courses were offered. Most schools supported a variety of athletic teams including basketball (both men's and women's), football, baseball, track, tennis, and golf.

The smaller camps either had no school and depended on the public school system or had small one room facilities with ungraded primaries. Sometimes, a single teacher would be responsible for the entire education of all the children who lived in a particular camp.

Commerce was usually conducted in company-owned commissaries. Most coal companies issued their own money called "scrip." This system worked much like an interest-free credit card. The miner could draw money in advance of his payday only on the amount of time he had worked up to the point of his request. He would be issued either coin or paper stamps which could only be spent at that company's store. The amount of the scrip issued would then be deducted from his paycheck. If he had money left over when his paycheck was due, he received the difference in cash. Some compa-

Located behind a railroad track, this Melcroft Coal Company Office and Commissary was located in Harlan County, Kentucky. May 16, 1923

Opposite the office and commissary at Melcroft Coal was the coal tipple where rail cars were loaded. May 16, 1923

Library of Congress/Farm Security Administration Project

"Coal Camp With Tipple And Dump Station"

nies were exploitive in this system by charging higher prices than other commercial establishments outside the camp or by paying only in scrip which required the miner to shop with the company-owned store. Many miners abused this privilege and ended up with no money at payday. This caused some of them to rationalize and blame the consequences of their behavior on the "evil" company.

As the economy began to change in the 1950s, these remote locations developed into prosperous communities. With alternatives to housing, mining companies were compelled to divest themselves of the residential housing and commercial retail businesses. Many decided to sell the homes to the miners at minimal cost and to close the commissaries. The scrip system also was discontinued. By the mid 1960s there were only a few company-owned towns still operating in the United States.

In the early 1900s, United States Steel used electric locomotives and steel cars to transport coal to dumping stations on the surface. May 1, 1919

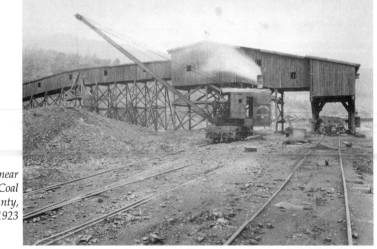

This steam shovel excavates near a wooden tipple at Melcroft Coal Company in Harlan County, Kentucky. May 16, 1923

This rare photograph shows a mule pulling wooden cars laden with timbers at a mining complex.

Mine Labor Unions

Coal miners have a rich union history. Since the early days of labor unions, coal miners have been active in organizing and promoting the value of strength in unity to accomplish better working conditions. Because mining is so dangerous and is usually conducted in fairly remote locations, living and labor conditions tend to be very difficult. The issues upon which miners have placed most importance include safety, health care, better pay, and fair labor practices. The three most significant unions in the history of coal mining in the United States are the United Mine Workers of America, National Miners' Union (1928), and the Progressive Mine Workers of America (1936). The largest of these is the U.M.W.A. which has had a history of strikes and protest to achieve wage benefits, better health care, improved working conditions, and enforceable safety rules and regulations.

John L. Lewis & the UMWA

John L. Lewis was the most famous and controversial of the presidents of the U.M.W.A. Both his paternal and maternal ancestors, the Lewises and Watkinses, migrated to the United States just after the American Civil War. His father worked in coal mines near Cedar Mines and Oswalt, Iowa. John was born February 12, 1880 at Cleveland, Iowa, and spent a brief time working in the mines in various capacities. His rise to the presidency of the U.M.W.A. was swift and unique. Just nineteen years after he was elected as a delegate and secretary of a newly formed local of the U.M.W.A., he was appointed President of the U.M.W.A.

As a young man, he was involved with the A. F. L. as a labor organizer but, throughout his career, had a tempestuous relationship with this powerful labor organization.

Benchmarks in the Life of John L. Lewis:

1897-1901 ~ Employed as miner in Lucas County, Iowa
1901 ~ Chosen as delegate and secretary of new local of U.M.W.A.
1906 ~ Chosen as delegate to U.M.W.A. Convention from Lucas, Iowa
1907 ~ Married Myrta Edith Bell, public school teacher
1908 ~ Moved to Panama, Illinois and went to work at Sandy Shoal Mine
1909 ~ Elected president of Panama Local 1475 of the U.M.W.A.
1911 ~ Appointed as an A.F.L. organizer for Ohio, Pennsylvania, and West Virginia

He and his beloved U.M.W.A. were principal in the development of the C. I. O. which made the issue of organized American labor prominent in the American industrial economy. According to Robert H. Zieger in his biography of Lewis *John L. Lewis: Labor Leader*, his lifetime dedication to the principle of unionism was largely responsible for the emergence of the United Auto Workers and the United Steel Workers.

His aggressive and commanding public speaking style garnered the attention of both followers and adversaries. His intelligence, ruthlessness, loyalty, and passion was both feared and admired by all with whom he came in contact. His forty-year leadership of the U.M.W.A. was littered with controversial principles of union strategy including militant organizing, antagonistic

Katherine Young

John L. Lewis served as President of U.M.W.A. for over forty years.

1916 ~ Defeated in a race for delegate to the 1917 A.F.L. Convention
1917 ~ Appointed statistician to the U.M.W.A. and business manager of the U.M.W.A. Journal
1917 ~ In October, appointed vice-president of U.M.W.A.
1918 ~ Elected to a full term as V.P. of U.M.W.A.
1920 ~ U.M.W.A. President Frank Hayes resigned John L. Lewis appointed to replace him.
1920 ~ Elected to a full term as President of U.M.W.A.
1921 ~ Defeated by Samuel Gompers for President of A.F.L.
1924 ~ Appointed to Advisory Committee of the Republican National Committee
1926 ~ Elected President of U.M.W.A. by defeating John Brophy

1930 ~ U.M.W.A. reorganized
1932 ~ Progressive Mine Workers of America formed
1933 ~ National Industrial Recovery Act Signed
1934 ~ Moved to Washington, D.C.
1935 ~ Committee for Industrial Organization (C.I.O.) meets for first time
1940 ~ Endorses Wendell Wilkie for President of the United States, but Franklin D. Roosevelt is re-elected.
1940 ~ Resigns as President of C.I.O. and is replaced by Phillip Murray
1946 ~ U.M.W.A. re-affiliates with the American Federation of Labor
1947 ~ Taft-Hartley Act is passed
1947 ~ U.M.W.A. disaffiliates with A.F.L.

1948 ~ W.A. "Tony" Boyle named as John L. Lewis' assistant
1948 ~ First U.M.W.A. pension check issued to a retired miner
1950 ~ B.C.O.A. formed (Bituminous Coal Operators' Association)
1955 ~ A.F.L. and C.I.O. are merged
1956 ~ U.M.W.A. hospital chain is dedicated
1960 ~ Retires as President of U.M.W.A.
1962-63 ~ Serves as Chairman of the National Coal Policy Conference
1963 ~ Tony Boyle is elected President of U.M.W.A.
1964 ~ U.M.W.A. hospital chain sold to Appalachian Regional Hospitals, Inc.
1969 ~ John L. Lewis dies in Alexandria, Virginia

Four adult miners pose with a young boy miner at a drift mouth at Lynch, Kentucky. 192?

relationships with the Federal government, and an aggressive authoritarian attitude.

John L. Lewis is best known for being a champion of the down-trodden coal miner but will also be remembered as an ardent capitalist who enjoyed the amenities associated with power and access to the elite of his day. Zieger characterizes Lewis as one of the great paradoxical figures of the 20th Century.

Women in Mining

Despite the common belief that women brought bad luck when they came underground at a coal mine, there is evidence of them having worked in the coal mines as slaves during the Civil War. A few worked underground in the early 1900s. At the turn-of-the-century, there is a report that some women in an Ohio mine worked disguised as men and were only discovered when they were killed in an accident.

According to Libby Lindsay, of the West Virginia Women Miner's Support Group, two sisters, Arna and Thelma Hubbard, worked when they were twelve and fourteen years-of-age at William's Mountain in Boone County, West Virginia, in the 1930s. There was an effort to recruit women to work underground during W. W. II and, despite opposition by the U.M.W.A., a few did hold jobs in scattered places throughout the coal fields.

According to Betty Jean Hall, former coal miner and now Administrative Law Judge in Virginia, the Civil Rights legislation signed by President Lyndon Baines Johnson in 1965 actually contained a clause which would have allowed women to work underground. But no one pushed the point until President Jimmy Carter signed executive order #11246 which reinforced the earlier legislation and prevented companies who had contracts with the government from discriminating against applicants because of race or gender.

The first documented case of a woman being hired by a commercial company was Debbie Pratt who worked for Bethlehem

The miner poses at a mine in Bell County, Kentucky. Female coal miners comprise only 2% of the mining work force. 1992

Steel Corporation K-4 Division in West Virginia beginning in 1973.

The number of women miners peaked in 1982 at 3,800. This still only represented approximately 2% of the total mining work force. Because of recent layoffs, the number dropped to 1200 in 1995. Since women were the most recently hired, they have been the first to be laid off.

Child Labor in the Mines

"Breaker boys, Woodward Coal Mines, Kingston, Pennsylvania." 1900

In the early days of mining, young boys were used to help in the coal industry. Some were called "breaker boys" and had the job of breaking the larger blocks of coal with sledge hammers or picking rock or slate from the coal as it exited the mine. Some served as "trappers." They opened and closed the doors to allow the coal cars to travel in and out of the mines. These doors had to be kept closed in order to direct the airflow to the working face. Many, as young as nine, labored in horrendous conditions to help their families earn a living. Some young men went to the mine with their fathers and helped them with their assigned working face. They were like apprentices but they were not paid. Later, when they gained experience, they were given a place of their own.

Child labor laws, passed in the early 1900s, prevented coal companies from further exploitation of these children in the mines.

"Breaker boys, Woodward Coal Mines, Kingston, Pennsylvania." 1900

"Breaker boys, Woodward Coal Mines, Kingston, Pennsylvania." Between 1890-1901

Historical Facts About Coal

- Native North American Indians fired pottery with coal-heated kiln-like devices in the eleventh century, long before Europeans came to the continent.

- Some of the native North American Indians in the east polished coal and used it for jewelry.

- Explorers and missionaries both reported that Indians used coal for trade and as a fuel.

- There were sightings of coal outcroppings in the 1600s as far west as Illinois.

- In the mid-1700s, coal was found throughout the mountainous regions of West Virginia, Virginia, Ohio, and Kentucky by hunters and explorers.

- Gunpowder was introduced to the mining of coal in the 1700s.

- 1713 ~ Thomas Newcomen introduced the use of steam to power an air pump which greatly advanced moving water from the mine site to the surface. This invention replaced the use of horse drawn pumps.

- 1745 ~ First commercial coal operation in the colonies began at Richmond, Virginia.

- 1760 ~ A group called "Penn Properties" purchased all the known bituminous coal fields in Pennsylvania from the chiefs of six Indian nations.

- 1770 ~ George Washington made an entry into his journal regarding a coal mine near Connellsville, Pennsylvania: "The coal seemed to be of the very best kind . . . burning freely and an abundance of it."

- 1771 ~ Bituminous coal was discovered in Richmond, Virginia.

- The American Revolution freed the country from dependence on British manufactured products. As Americans began to establish their own industries, coal became a critical element in that process.

- 1780 ~ Wooden and cast iron rails are replaced by the introduction of wrought iron.

- 1784 ~ Coal was first mined near Pittsburgh for blacksmithing and space heating.

- 1815 ~ After use of animals such as dogs and canaries to detect the presence of deadly gas, Sir Humphrey Davy invented the Davy Safety Lamp which worked on the theory that flame will not pass through a fine wire gauze if its apertures are a certain size and thickness.

- 1866 ~ After years of commercial surface mining being done using picks, shovels, and wheelbarrows, the first strip-mining occurred using horse-drawn plows and scrapers at the Grape Creek Mine near Danville, Illinois.

- 1877 ~ An Otis-type steam shovel was used near Pittsburg, Kansas, to clear twelve feet of overburden for a three foot seam of coal.

- 1885 ~ Consolidated Coal of St. Louis, used a converted wooden dredge to strip coal at Mission Field.

A wooden coal car sits in the main drift mouth of Mine #30 at Lynch, Kentucky, July 1, 1920.

49

Mule drawn wagons being loaded by coal fired steam shovel, Lynch, Kentucky. 1920

- 1889 ~ The Butler Brothers used the first draglines to mine coal at Mission Field, Idaho.

- 1893 ~ Pennsylvania State College began extension classes in mining.

- 1910 ~ 710,000 coal miners dug 417 million tons of bituminous coal and 85 million tons of anthracite.

- 1911 ~ Grant Holmes and W. G. Hartshorn designed a long-boomed, self-propelled shovel capable of digging and dumping in any direction. A machine builder named Marion built the machine which had a 3 1/2 cubic-yard dipper on a sixty-five foot boom with a forty-foot dipper stick.

- 1912 ~ Bucyrus Company purchased Marion shovel and produced an improved steam-powered shovel.

- 1930 ~ Walking draglines were developed. These machines walked on hydraulic legs which lifted the machine, moved it for-

ward, and set it down again.

- 1932 ~ 5,427 mines were in operation in the United States, employing 400,380 miners.

- 1933 ~ 10% of coal in the U. S. was mined by machine.

- At the start of World War II, 630 million tons of coal were produced in the United States.

- 1949 ~ *Deep Mining*, first detailed study of mining problems, was published by Jack Spalding.

- 1953 ~ 80% of coal in the U. S. was mined by machine.

Mine rescue teams were sponsored
and trained by coal companies to
have a quick response to mining
accidents.
This team poses at Benham,
Kentucky.
1960s.

Miner and wife with mule
and wagon 1960s.

Miners kept their clothes on hangers pulled to the ceiling of the company bath house where they could dry in the steam heat. May 1, 1920

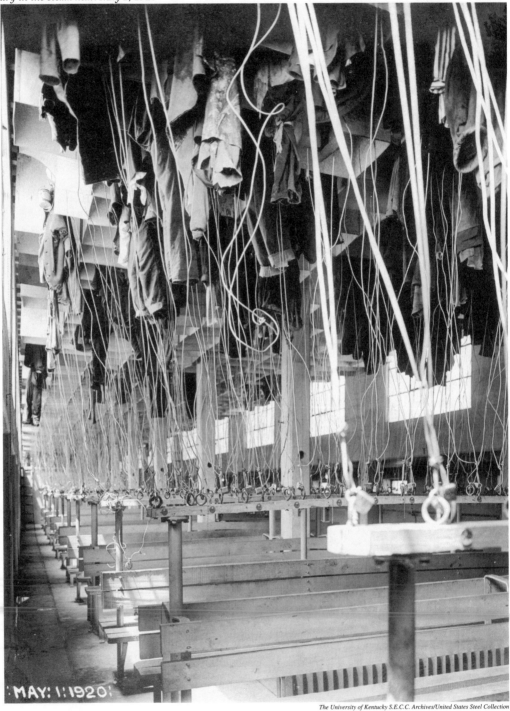

MAY: 1: 1920:

Coal Miners' Dress & Personal Equipment

Coal miners dress specially for the conditions under which they will work. Underground miners are required to wear an O.S.H.A.-approved hard hat, safety-toed shoes or boots, a self-rescuer, and safety glasses. Additionally, elastic bands are required around the bottom of their pant legs to keep them from being caught in the moving parts of machines. Miners often decorate their hard hats with stickers which advertise various mining equipment and supply companies. They also affix their names or nicknames.

Clothing usually consists of overalls, coveralls, or just plain heavy cotton work shirts and pants, a mining belt (nylon or leather), heavy cotton or wool socks, and gloves.

In addition to minor repair tools, the miner might be equipped with various other kinds of personal items: knee pads, a dust sampler, an anemometer, ear protection, a noise sound instrument, a safety lamp, a dust respirator, and a sounding rod.

Most underground miners get dressed in a company-provided bathhouse. This is a building equipped with showers and a "dry room" where miners can hang their clothes for drying. The clothes are placed on a device equipped with hooks which is pulled to the ceiling by chain and pulley. Additionally, there are places for the miner to put his personal belongings and toiletry items. Each of these hangers is secured by a padlock.

Coal miners of the 1920s had some variance in their work clothes--some wore jean jackets and some wore wool suit coats. 192?

The University of Kentucky S.E.C.C. Archives/International Harvester Collection

The University of Kentucky S.E.C.C. Archives/International Harvester Collection

*These miners are operating a chainsaw-like device
called a "cutting machine" which undercut the coal
and prepared it for blasting. March 25, 1919*

Two Types of Coal Mines

Coal mines are sometimes grouped by where the coal is deposited in the ground. Sometimes the coal lies close to the surface and sometimes it is buried deeply within the Earth. The two ways coal can be reached is by underground or surface mining.

Many modern mines use a meat slicer-like device called a longwall miner. This one was at a U.S.S. mine near Lynch, Kentucky. 198?

James B. Goode, Photographer

Underground Mines

Three Types of Underground Mines

Underground mines can be grouped by how the coal mine owner elects to reach the coal that is buried inside the Earth:

1) **A Drift Mine** is used when the coal outcrops or is close to the surface. This coal usually lies horizontally (or nearly so) and can be reached or exposed from the side of the hill. The opening is made directly into the coal seam. Coal is transported by rail cars, self-powered vehicles, or conveyors. Miners, equipment, and supplies are transported by rail cars, self-powered vehicles, or conveyors.

These mines can extend for miles underground. This is an economical mining technique because it rarely requires excavation through rock.

2) **A Slope Mine** is used when the coal does not outcrop and must be reached from a perceptible angle. An inclined tunnel is dug through rock strata to reach the coal bed. Machinery used in this type can either run on its own power or be hoisted by cables. Miners, equipment, or supplies are transported by rail cars or by self-powered vehicles. Coal is moved to the surface either by rail car or slope conveyor belts.

These mines can extend for miles underground. Slope methods are frequently used in conjunction with shaft methods.

3) **A Deep-Shaft Mine** is used when the coal must be reached deep beneath the Earth's surface. In this type, two or more vertical shafts are sunk to the depth of the deposit: one houses the elevator that will haul coal to the surface and transport miners and equipment in and out of the mine; the other provides ventilation. A large fan at the top of the ventilation shaft brings fresh air to the miners and pulls away stale air and dangerous gases. The elevator is operated by "winds" which are powerful engines which operate huge drums equipped with steel cable. These are usually located in a building adjacent to the actual shaft and are similar to the wenches used by wreckers and four-wheel-drive vehicles. The depth of these shafts average 300 feet in the United States and over 800 feet in Great Britain.

The Miner's Lunch Pail - The top removable section holds food while underneath there is a compartment for liquids.

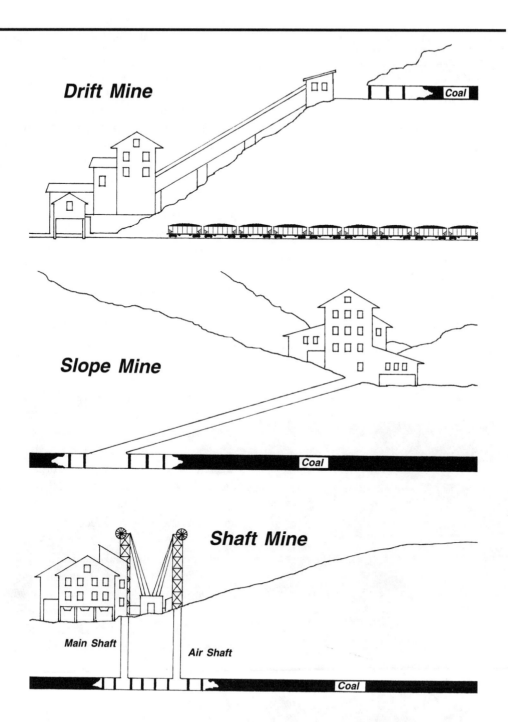

Drift Mine

Coal

Slope Mine

Coal

Shaft Mine

Main Shaft

Air Shaft

Coal

Two Underground Mining Systems

Sometimes mining is described by the system used to remove the coal. There are two basic systems:

1) Room and Pillar

In this method, only part of the coal is removed. Pillars of coal are left to support the roof and prevent cave-ins. The average coal recovery rate using this method is 50%. If the operator robs the pillars, coal recovery rates increase to 80-85%. The specific mining plan varies with local conditions encountered. The general procedure is to drive main entries along the seam from one or both sides. Side entries are then cut, forming a cross-hatch of tunnels at right angles. As coal is removed, the side entries become large rooms. Entries are spaced far enough apart to leave a pillar of coal between rooms of sufficient size to adequately support the roof. The size of the rooms and pillars depends primarily on the character of the coal and the amount of roof pressure.

The mine is ventilated by forcing air down the main entry by a huge electrically operated fan. The air is then distributed throughout the mine by various concrete block walls called "stoppings" or temporary plastic curtains called "brattices." Air is then exhausted through an auxiliary entry. The entry housing the conveyor belt is neutral for airflow which prevents miners from working in an environment of smoke if there is a mine fire.

2) Longwall Mining

This method was borrowed from Europe and introduced in the United States in 1966. The coal is removed from a large face ranging from 300-650 feet by a shearing machine which planes back and forth across the face allowing the roof to gently cave in behind it. The advantage of this method is an astounding 85% coal recovery rate completed in one operation. This method eliminates roof bolting and creates a 15-20% reduction in manpower. Longwall mining takes less supervision, and simplifies ventilation service operations.

Sometimes this method is the only one suitable under weak roof conditions.

James B. Goode, Photographer

Irvin Turner operating remote control miner at Underground Mine #37 near Cumberland, Kentucky. March 6, 1995

Underground Mining Methods

Another way to describe coal mines is to talk about the way the coal is mined in them. Sometimes the coal is drilled and blasted, sheared or ripped, or picked and shoveled.

In Conventional Mining, a cutting machine is brought to the face (the vertical wall at the end of a corridor). A deep slot is cut in the coal at the bottom of the face which will help the coal fall and shatter when it is blasted. A mobile drill is moved into position. Holes are then bored and spark-proof explosives or cylinders of compressed air are loaded. The machine is then removed. The charge is detonated and the coal falls to the floor. A loading machine is brought in and the coal is swept onto a chain conveyor which loads a shuttle car. This rubber tire car then transports it to a conveyor belt which carries the coal outside. After the coal is removed, a roof bolting machine is moved into place and the roof is drilled and bolted. The bolting of the roof holds the rock strata and helps prevent potentially dangerous rock falls.

Continuous Mining was introduced in the 1950s and involves a machine with a rotating cutting head tearing the coal loose in one operation. The continuous mining machine contains its own loading device which moves the coal to the shuttle car. The yield by this method usually varies from 200-400 tons per crew shift.

James B. Goode, Photographer

This longwall shear works a twelve foot seam of bituminous coal at a U.S.S. mine near Cumberland, Kentucky. 198?

My Trip to Arch of Kentucky's # 37 Underground Mine

Bob Lunsford and I were the veterans of the group invited to tour the underground mining operation of Arch of Kentucky's #37 mine on Cloverlick Creek in Harlan County near Cumberland, Kentucky. The other two, Stephanie Peace and Lori Chambers, had never been underground. Stephanie was an Appalachian Studies student at Southeast Community College and Lori was a reporter for the Harlan Daily Enterprise.

We met in my office at 7:30 a.m. I had already dressed for the occasion. I had on coveralls which had the bottom of the legs secured tightly with elastic bands, a mining hard-hat, steel-toed boots, safety glasses, and my leather mining belt with a metal I. D. tag riveted to the outside. We drove my Pathfinder up Cloverlick Creek to the mine location. We gathered our equipment and headed toward the blue metal building which houses the bath house and mine offices. A member of the Safety Department greeted and escorted us to an upstairs classroom for Federally required safety training. Every miner must have extensive safety training periodically and every visitor must be given a short course in mining safety. We were told about the way the mine was laid out, how it was ventilated, where the main passages were, and the location of the conveyor belt lines. Instructions were given about what to do in case of an emergency such as a mine fire, roof fall or explosion. The instructor demonstrated the use of a self-rescuer which filters harmful gases such as carbon monoxide and is carried on every miner's belt. He explained that these self-rescuers do not supply oxygen but that there are self-contained oxygen apparatuses located in designated stations throughout the mines.

After he answered a few questions, we were told that there was a canvas bag under each of our chairs which contained all the personal clothing and equipment we would be needing underground. As the crew donned their clothing, I asked the trainer about the kind of photographs I would be allowed to take. Some coal companies are very particular about the kind of images they allow photographers to use. The trainer told me that their policy was to allow open access to the mining operation.

We went back downstairs to the lamp shop. The company employs a man who stores, maintains, and repairs all the wet cell electric lamps that are issued to each miner everyday. He issued a light to us. The lamps have rectangular plastic batteries, about the size of a medium sized textbook, which have an insulated 3' cord attached and an encased, sealed bulb at the other end. We mounted the battery on our belts and clipped the encased bulb to the front of our hard-hats.

The mine superintendent in charge of the shift we were visiting was Lonnie Lowe. Lonnie met us in the classroom and then escorted the group outside to the electrically operated man trip. This is a small train of personnel carrier cars pulled by an electric motor or locomotive. The unit was backed under a metal shed and waiting for us to

Our tour guide, Lonnie Lowe, poses beside an eight-ton longwall mining jack at an Arch of Kentucky mine near Cumberland, Kentucky. 1995

board. The carriers we used were low, covered with steel canopies, and had plastic curtains over the doors to deflect the wind. The mines are very cool year around. The average temperature is approximately 50-55 degrees. But because of the tremendous circulation of air by huge fans, the chill factor can be very low in the winter.

These transport cars have two facing bench seats and will carry six miners comfortably.

With Lonnie at the controls, we started with a jerk and the electric motor whined as we made our way toward the gaping black hole in the side of the mountain. As we approached the drift mouth, we passed several men performing various tasks outside the mine. Some were loading conveyor belt onto a truck. Others were working on equipment. A few were shoveling around the structure where the conveyor belt came out of another hole in the mountain.

As we entered the mine, I carefully leaned out of the car to look back at the opening. It grew smaller and smaller, until it was a tiny dot. Finally it disappeared altogether. We all switched on our wet cell battery lights. We were cautioned that when we looked toward someone's face, to direct the light just slightly to the left or right of their ears—that would prevent the light from blinding them. Lonnie warned me to keep my head in the car. He explained that there are rocks and other obstacles which might protrude from the rib. These could strike me and cause injury or death. He warned that we would be passing other pieces of equipment which might come dangerously close.

The track curved and dipped. We passed several switching points where the track split and went into another section of the mine. As we directed our light to the roof of the mine, we could see the grooves cut by the carbide tipped bits of the continuous miner.

Occasionally, we caught the glimpse of an ancient fossil fern or the outline of a tree limb or stump. At one point, we passed some men laying concrete block across an opening in the side of the tunnel. Lonnie explained that this was a "brattice" or "stopping." These walls are used to direct the airflow to more active sections of the mine.

We finally pulled up behind a stopped

supply car. A crew of miners was unloading bags of rock dust. Lonnie opened a small place on the end of one bag and took a handful of the talcum-like dust.

"This is used to reduce the likelihood of a dust explosion," he said. "Rock dust is inert and will not ignite. We either scatter this over the rib and bottom in a dry form or mix it with water and create a slurry which is sprayed over the same area."

We all ran our fingers through the silky, cool, powder.

"When an explosion occurs, from igniting methane gas, coal dust is picked up from the floor of the mine by the force and turbulence. As the coal dust is suspended, the fire ignites it and causes more turbulence. This causes even more coal dust to be suspended and pretty soon you have a storm very much like a tornado, except it's full of fire."

We looked cautiously at the coal dust lying on the bottom.

"When a miner is caught in one of these tunnels during an explosion it's like being inside a gun barrel," he continued. "I've seen coal cars blown clean across the hollow from the fierceness of the explosion."

"How do you prevent this ignition from happening?" one of the girls asked.

"We do not allow any smoking in the mines. Miners are not allowed to have matches, cigarette lighters, or even cigarettes. All smoking material is banned. We also do not allow any equipment that is not permissible. 'Permissible' means that the Mine Safety and Health Administration has ap-

proved the piece of equipment for underground use."

"In the British Isles, miners are not even allowed to wear watches or hearing aids which are battery operated," I offered. I had spent some time touring Welsh coal mines during a trip a few years before.

We walked down a long, black corridor toward a loud noise which sounded like the air tools used at tire shops. There was also a thunderous rumble. We could see some miner's cap lights in the distance. As we approached, we met the face boss standing in the roadway just outside a room where a large, noisy, continuous miner chewed away at the coal seam.

Another man stood just to the left of the machine with a rectangular box hanging from his waist which contained small joysticks and buttons. He was operating the giant machine by remote control. The front of the continuous miner looked like a huge paint roller equipped with hundreds of cylindrical metal carbide-tipped bits mounted all over it. The roller turned and turned as the operator raised and lowered it with his controls.

A shuttle car operator had pulled his rubber-tired carrier under the conveyor which protruded from the back of the mining machine and the coal poured into the bin in the center of his car. A short burst of the electric drive motor moved a chain conveyor in the car and the coal back toward the rear of the shuttle car. We had to yell over the tremendous noise.

The face boss, or man in charge of the mining at this particular face, explained that

Mine Foreman Paul Royce poses during a busy shift at Arch of Kentucky's mine #37 near Cumberland, Kentucky. 1995

James B. Goode, Photographer

they were cutting the 500' X 4,000' blocks that would eventually be mined by the longwall machine. We retreated to a quieter place called a "break" or a doorway between rooms where we could talk.

"What is a longwall miner?" Lori asked.

"A longwall mining machine is a giant mechanical device that operates much like a meat slicer at a grocery store. We set the machine on the 500' side of the block of coal called a 'panel.' The machine has two arms which are equipped with cylindrical drums called 'shears.' These are fitted with bits much like the continuous miner. The only difference is that the shears are much larger in diameter and are pulled along by a giant chain. The coal is cut as the shear travels either direction. As the shear is set into the coal it is turned by a powerful electric motor which causes the bits to 'shear' the coal from the face. The width varies depending upon the machine, but the one we are looking at is 48" wide," Lonnie explained.

"How does the machine move once it's made a pass?" I asked.

"As the shear turns and is pulled into the seam, the coal falls into a large chain conveyor, is crushed, and carried out of the mine on belt conveyors," he continued.

"The machine contains dozens of vertical and horizontal hydraulic jacks which are connected to big metal panels overhead and on the bottom. As the shear passes along the face, the machine advances, like a big, side-winding caterpillar, into the coal."

We walked back out into the roadway and made our way toward the bright lights farther down the tunnel. When we arrived the shear was making a return pass. The dust was so thick I could barely see the drum. Coal fell onto the chain conveyor in chunks as big as a refrigerator. The giant machine slowly crawled toward the coal seam and then snuggled next to it.

The men on the section wore special helmets shaped like large, curved footballs. The helmets had face shields and were designed so the back protruded downward to protect the miners' necks from falling debris. They were also equipped with small, circulating fans which cooled the miners' heads.

We stayed a short while, watching the advancing beast. We could hear the top falling in behind the machine. The men were working inside an advancing cocoon. They were in the belly of the machine—safe from the tremendous, crushing weight from overhead.

Lonnie motioned for us to exit the machine. We made a hasty retreat to a quieter part of the mine. We stopped to watch some men drill the roof and place long metal rods called roofbolts into the holes. These were "glue bolts" and the men placed the resin glue contained inside plastic tubes into the holes, pushed the bolts equipped with 4" square metal plates as collars, into the holes, spun them with a powerful electric motor, and moved to the next area.

"These bolts will prevent most of the rock from falling to the bottom," he said.

"Our roof control plan also calls for the use of mine props or timbers. We place posts made from trees in various places in the mines to assist the roofbolts. Sometimes we have to build a cage of 4" X 6" square pieces of wood 2' long called "cribs" to help hold the top." Lonnie pointed to a tall crib just to our right.

We made our way back to the man trip car. We all sat in the cool, damp car as Lonnie goosed the electric motor and pulled onto the main track. We didn't say much on the way out. I guess we had been mesmerized by the experience. After we turned in our gear and said good-bye to Lonnie, we made our way to the car. On the trip back to town, Lori chattered incessantly. The technology had not been what Lori had expected.

Being a coal miner took more training than she had first thought. She couldn't believe the amount of coal being mined by the longwall machine.

Stephanie had encountered her dad, who is a section boss, during the trip. She beamed proudly as she recounted his explanation of how his methane monitor could detect the explosive gas and warn him if the concentrations were too high. She was proud that he had passed the state and federal tests required to become a foreman in a coal mine.

Bob Lunsford and I, two veteran visitors to the mines, smiled knowingly. I looked at them all and said, "I am always amazed that coal miners spend everyday stepping into frontiers where no man has ever been."

A more recent version of the miner's helmet

*This miner shovels coal into the bucket
of a machine called a scoop.*

James B. Goode, Photographer

David Duff, owner of Pine Branch Coal Sales, looks at his
Combs' Branch Red Rock #10 strip mine near Chavies,
Kentucky. April 29, 1995.   *James B. Goode, Photographer*

Strip or Surface Mines

Four Types Of Strip Mining

1) Area

In the area strip mining method, the land utilized is fairly flat. This technique involves removing the topsoil, stockpiling it so it can be used later, and then dislodging and storing the earth and rocks called "overburden" by using scrapers and/or bulldozers. The overburden is usually dislodged by blasting or scraping. The coal is usually blasted with a mixture of ammonium nitrate and fuel oil and scooped up by claw equipped machinery such as shovels, scrapers, or end-loaders. Sometimes the coal is soft enough that blasting is not required and it can simply be scooped up and loaded.

The coal is then hauled away by trucks to loading tipples. The overburden is then replaced, graded to original contour, topsoil is returned, and the site is replanted with vegetation.

Equipment ranges from ordinary bulldozers and front-end loaders to immense power shovels and draglines. Many of these machines are the largest mobile machines in the world and have to be assembled on site.

2) Contour

In the contour strip mining method, the topsoil is removed and stockpiled for reclamation purposes. A bench is bulldozed into the side of the slope, the rock is drilled and loaded with explosive, and blasting loosens the dense overburden. The loose material covering the area is scraped, scooped, and hauled away; the coal is removed by loaders and/or shovels and is transported out along a haul road built especially for this purpose; the area where the coal has been removed is back-filled with spoil material consisting of rock and sub-soil; and while blasting for the next stage of overburden removal goes on, reclamation of the first cut is beginning. The pit is filled with overburden, re-graded, layered with topsoil, then seeded.

3) Mountain Top

In the mountain top removal strip mining method a drill bench is cut from the side of a mountain, both for use as a haul road and for extending drilling. As in the other two methods, the topsoil is removed and stockpiled, the overburden is drilled for placement of explosives, and blasting loosens the overburden. Loaders or shovels load the overburden into trucks and it is used as back-fill in a previously-mined portion of the pit or placed in a head-of-hollow fill. The exposed coal may be blasted or loaded from the seam, depending on its hardness. Trucks haul the coal out of the pit area. The back-filled pit is graded, spread with topsoil, and revegetated, while the next "cut" is begun. A flat to gently-rolling area results.

One-fifth of all coal is suitable for strip mining because it is within 100-150 feet of the surface. When the overburden is in this range, it is often too fragmented and weak to form a safe roof for an underground mine. This method has accelerated due to rapid increases in underground costs. These sites

*Excavator loads overburden into truck
at Pine Branch Coal Sales' Lead Branch strip mine
near Chavies, Kentucky. April 29, 1995*

Augering is not very practical when one considers that only 50% of the coal is mined and the rest will probably never be recovered.

As the drill completes its task, the coal is loaded by conveyor and trucked out of the mine area. Then reclamation of the pit begins. The holes are covered with spoil, the overburden is returned and graded, topsoil is restored, and vegetation is established.

can be opened faster, have fewer labor difficulties, are safer to operate, and have an 80-90% recovery rate. The average production-per-man is three times that of underground mining.

4) Auger

The auger strip mining method is a supplementary mining method used to recover coal which has too much overburden to strip economically. This method was introduced in 1945 and is responsible for 3% of total production in the United States. After the coal has been mined out to the desired distance into the mountain, an auger or drill measuring up to 60" in diameter is used to bore horizontally into the seam and perpendicular to the bench. Auger depths can be as much as 250 feet.

As the auger bores, it carries the loosened coal back out to the pit area. When the first bit has penetrated to its maximum length, another bit is attached to it and the drilling starts again. Some coal is left between the holes to support the mountain and reduce the amount of surface subsidence.

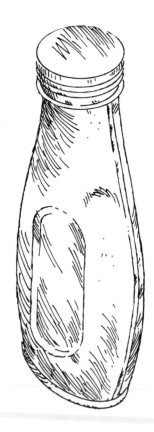

This metal container is for an extra supply of carbide

An end-loader dumps coal into a transport truck at Lead Branch strip mine near Chavies, Kentucky. April 29, 1995

David Duff, owner of Pine Branch Coal Sales, looks on as a bulldozer clears the roadway at his Lead Branch strip mine near Chavies, Kentucky. April 29, 1995

My Trip to Pine Branch Coal Sales
April 29, 1995

My visit to Pine Branch Coal Sales at Chavies, Kentucky, was very interesting and informative. The Duffs have been in the mining business for the past twenty years. They have had both deep mines and strip mines. Today, they are primarily in the strip mining business. I discovered that the techniques and requirements for strip mines or surface mines are vastly different from those for drift or shaft mines. Strip mining involves removing the "overburden" or trees, soil, and rock so that the coal seam is exposed. The coal is then scooped up and loaded into trucks which carry it to "loadouts" or tipples.

The trees are often cut and used for lumber or deep-mine props. An advance crew with chain saws and logging equipment clears the trees and brush. The Duffs have used lumber sawed from their mine property to build houses and offices. Some of it has been sold for profit to other businesses.

I spent the night with David and his wife Susan in their home on a beautifully landscaped hill near Chavies, Kentucky. David and I arose at 6 a.m. and put on our khaki work clothes. David put on his low-rider, antique, black miner's hard-hat. He clipped on two Motorola Radios and a cellular phone. We picked up a rather large cooler from the kitchen and placed it in the bed of his four-wheel-drive truck and were on our way shortly before 6:30 a.m.

Almost before we started the truck, David was on the radio talking to someone with the handle "Doggy Dog" who was already on the strip job.

"10-4, Doggy Dog. . . I'll be up there stat to check the jack on that 992." David keyed the microphone clipped to his shirt pocket.

We went through a security gate operated by remote control. David punched the accelerator as we gained speed in a long, wide, straight stretch of gravel road. We passed a truck filling up with water from a large reservoir tank on the hill.

"We have a reservoir pond just to the right there." He pointed to a large pond with beautiful grass growing along its banks. "We pipe water from several sources back in the hollows and then, with a pump I specially designed, move the water to the holding tank. We use the water to keep the dust down on the roadways. The man you see is full-time and does nothing but fill and dump water during his shift."

"Does the water serve any other purpose?" I asked.

"Well, you wouldn't believe how much money it saves on air filters for our equipment. They're very expensive and the cleaner we can keep the air, the less it costs us in maintenance." David was driving rather fast on the wrong side of the road.

"Aren't you on the wrong side of the road?" I asked nervously.

David gave a big horse laugh, finished off with a snort. "Our operators are trained to pass on the narrow roads. We pass on the left because of the possibility of collision. The driver is then farthest away from the

Early morning fog rests on Pine Branch Coal Sales' Combs' Branch Red Rock #10 strip mine near Chavies, Kentucky. April 29, 1995.

James B. Goode, Photographer

potential impact and in less danger of being hurt."

We continued to wind up the mountain. David never slowed down.

"Aren't you afraid you'll hit a rock in the road while you're traveling this speed?" I needed reassurance.

"The bulldozer and road scraper operator build roadways and keep them clear. Ditches are also built to divert water as it runs off after hard rains. We really can't afford to allow rocks to stay in the road. The 994 Caterpillar's tires currently cost $29,000 each. The front tires only last for about 2,000 hours of operation or 3 1/2 to 4 months. So you can see, we can't afford to have many blowouts!"

As we pulled into one of the pits and exited the pickup, the radio crackled again. "We're gonna need someone up here to help finish ramping down in the Lead Branch operation. Can you work the Dll-N?"

"10-4 Jay. . . I'll hit that soon as I get to this jack." David walked toward two men standing next to a giant end-loader which had its bucket propped on the back of an equally giant dump truck.

"What kind of truck is that?" I was awe-struck by its size.

"The equipment we use in removing the overburden and coal is very large. Our family primarily uses equipment made by the Caterpillar Corporation. The big dump trucks are Caterpillar 777Bs but we also have 773s, 785s, and 789Bs. The 777B hauls 85 tons and costs between $900,000 and $1 million. The 785 holds 150 tons and costs $1.2 million each. The 789B is twenty-five feet wide, holds 190 tons, and costs $1.4 million. The 789Bs tires are $14,550 each. The largest of these trucks weighs more than six 737 jet airplanes." David amazed me by being able to rattle off these statistics without notes.

"The end-loader we're working on here is a Caterpillar 992C but we also use a 994. The 992C costs between $950,000 and $1 million, holds 13 1/2 cubic yards in its bucket and requires six buckets to fill a truck. The 994 costs $2 million, has a capacity of 23 cubic yards and requires about 4 buckets to fill a truck. In a typical ten-hour shift, the loaders can move between 22,000 and 27,000 tons of rock. The twelve-hour night shift moves between 28,000 and 34,000 tons of material." He looked at the machine schematic, continued to talk to me, converse on the radio, and discuss the repair with the two men working on the end-loader. Here was a veritable "quadrasonic" man.

The problem with the hydraulic jack which raises and lowers the bucket of the end-loader was soon solved and David headed the truck toward the Lead Branch section of the mine.

When we approached the pit David took a detour and we went up a very narrow, steep dirt road which led to the top of the high wall. He wheeled the truck around and backed it into a narrow space between two rocks. He bounded out of the truck, grabbed the cooler and headed to the edge of the high wall. He opened the cooler. Inside was the largest lunch I have ever seen. Susan had packed at least a dozen sandwiches of various kinds. She had potato chips, corn chips, pickles, soft drinks, and two of the biggest bunches of green onions I've ever seen. I had heard that the Duffs are legendary in their love of onions!

As we ate lunch we watched the end-loader scoop the rock and dirt and load it into the big truck. We watched the big bulldozers making ditches and keeping the roadway cleared. I asked David to talk about his bulldozers.

He chewed on an onion as he began talking, "The bulldozer you see is a Caterpillar D9-H; we also use the D8-K, D10-N, and D11-N. The blades are designed to move both dirt and rocks. The D-11N is the biggest machine and costs $900,000. When the bulldozer tracks wear out, the replacements can cost over $20,000 each. An entire undercarriage overhaul, which includes rollers, idlers, segments, and tracks can cost as much as $71,000!"

As I watched the smooth rhythm of the end-loader and dump trucks, I asked David about the skill involved.

"Our drivers are highly trained and experienced. The youngest on the job has been here ten years. The drivers must also know how to approach the end-loaders to make each trip most efficient. The end-loader operator clears and piles loose debris if he has any extra time while waiting for the big trucks to return," he explained.

"Matching the right equipment is necessary to efficient mining. For example, it is important to match the right size end-loader with the right size truck. Having two different size trucks being filled by the same loader will cause wasted movements or prolonged waiting. This can cost me thousands of dollars. The way trucks are loaded involves a lot of planning. Communications and rhythm are equally important. All the equipment has up-to-date radios and operators are able to talk, even over the roar of the big engines."

"Explain exactly what I'm seeing from up here. . ." I asked.

"Heavy equipment such as bulldozers, backhoes, and end-loaders are used to remove the soil down to the bedrock. Topsoil is taken to a holding location nearby and stored. It will be used later in the reclamation process. When we reach the bedrock, drilling machines are used to bore holes deep into the strata. These are then loaded with explosives and wired to an electronic detonator. The charges are carefully placed so that the explosion "casts" the rock toward an area which has been cleared for this purpose. The coal is then exposed so it can be removed. The drills I use are Ingersol/Rand DM 45s, have 45,000 lbs of pull, and bore 7 7/8" holes up to 130' deep."

David walked to the edge of the high wall to talk to someone on the radio. As I looked down, I saw a crack developing in the dirt under his feet. It looked as if it went several feet down.

I started to warn him but before I could say anything he said, "I have to go down and get on the dozer. You can just hang out and take pictures—I won't be long."

We took the truck back down the hill and David vaulted onto the bulldozer seat. He appeared perfectly comfortable operating the big machine. While he was working, I walked over to the bucket of a big end-loader where one of the operators and a truck driver were having lunch. They explained that David was a "hands-on" operator.

"He can operate any piece of equipment on this job," Red Hubbard, the end-loader operator declared.

"I've never seen anybody more talented. He can also repair any piece of equipment up here. If you call David Duff and ask him for a part number, he can give it to you nine times out of ten, without looking it up!" Ben Smith, the truck driver added.

I shot a few pictures and, after a few minutes, David finished his bulldozer work. He motioned for me to come to the truck. I walked through the fresh smelling, red earth toward his truck.

"Let's go back on top of the high wall, I've got to talk to someone on the radio and need a clear shot to the repeater," he explained.

We flew up the steep incline. Just as we reached the top, I saw David jerk the steering wheel sharply to the right, open the

Strip miners at Pine Branch Coal Sales near Chavies, Kentucky eat their lunch in the bucket of a giant caterpillar end-loader. April 29, 1995

driver's side door quickly, and slam it shut. He cut the truck quickly to the right. When we got out, the truck was sitting on a small mound of dirt and everything else was gone! The entire place where David had been standing, not thirty minutes before, had sheared off into the pit. All that was left was our parking space. There was huge V cut into our access road.

"Dangit! Dadburnit!" David kicked the dirt with the toe of the boot. You'll have to come with me."

We started walking back into the woods.

"Dangit!" David kept saying. Finally, we approached the biggest backhoe I had ever seen parked on the edge of the woods. "This is a 245 Caterpillar backhoe/excavator," David beamed proudly. It has a capacity of 3.5 cubic yards per bucket."

I stood next to him, holding on for dear life on a vertical piece of metal rod mounted on the cab as he made his way toward the slide area.

"I've got to build us a road!" he explained. He moved the excavator into position and began moving dirt and rocks over the high wall. Soon a primitive road appeared. We parked the 245 and took the truck down the precarious path to the floor of the pit.

The day was getting late and David wanted to take me to a drill site. We drove over to where a big Ingersoll-Rand Drill moved along on steel tracks.

David explained the process as I

watched the crew work. "The blasting process at Pine Branch involves several steps: 1) The rock is uncovered and its depth is estimated; 2) the holes are drilled to the approximate depth. They are drilled 18' apart in rows that are 23' apart; 3) the holes are measured a second time to make sure they are the same depth; 4) a detonator called a 'cap' is attached to a primer cord and lowered into the hole. At Pine Branch, the primer cord is non-electrical and consists of a flexible, hollow plastic tube, similar to a soda straw but much smaller, which is filled with a fine line of black powder; 5) a truck loaded with a mixture of ammonium nitrate and fuel oil fills each hole to within one foot or so of the top and the remainder is filled with the drill chips called 'stemmings' which is dust created by drilling the rock; 6) the primer cords are connected on the surface; 7) the primer cord is rated in various milliseconds so that each hole explodes exactly when we want it to. A device, similar to a firecracker, ignites the powder inside the tubing and an explosion occurs in each hole in a sequence planned to shatter the rock in a certain way. Some of the ignitions in primer cord can travel up to 14,000 feet per second."

After we had watched the drill operation a few minutes, David drove to a site where reclamation had been completed. Cattle grazed peacefully on a gently sloping hillside. The place looked very much like land I had seen during a stay in the coal mining areas of Wales a few years before.

"My Dad, Don, is in charge of the cattle operation. We have hired a rancher from Texas named Larry Clay to help him manage the ranch. The family calls him 'Cowboy.' We have a very successful beef cattle business here and market several hundred head per year."

"How many mines do you own and how much coal do you produce?" I asked as we headed back toward the peaceful, grassy hillside where David's and Susan's house is located.

David took his hard-hat off and scratched his thinning hair. "Pine Branch Coal Company currently has three mining locations and produces an average of 130,000 to 140,000 tons every 25 days. Our record is 160,000 tons in a 25 day period. This amounts to almost 1.6 million tons per year."

As we barreled down the gravel road, I contemplated those statistics along with everything I'd seen that day. This was truly an eye-opening experience. All of a sudden, David slammed on the brake and said, "Dangit!" again. He jumped out of the truck and ran back up the road. As I watched him, he bent over to pick up a paper plate someone had thrown on the roadway. He carefully picked it up, folded it, placed it in his pocket, and returned to the truck.

"I'll not tolerate anyone littering my roadway," he said while clenching his jaw. "I won't put up with it."

Arch Coal Strip Mining

Facts About Strip Mining

■ Strip mining uses the largest of all mining equipment and involves the greatest expanses of land.

■ Surface mines are 3 times as productive as underground ones and involve fewer potential hazards for miners.

■ Almost 100 percent of the coal at a strip mine is recoverable.

■ One fifth of all coal in the U. S. is suitable for strip mining because it is within 100-150 feet of the surface.

■ Average production per person in strip mining is three times that in underground mining.

■ In 1975, for the first time in history, surface mining fatalities passed those for underground mining.

Arch Coal, Inc.

Arch Coal Reclamation

The largest coal tipple, for its time, was built at United States Steel's mines at Lynch, Kentucky. September 1, 1920

How Coal is Processed

Once the coal exits the mine there are several essential elements to processing the mineral before it can be used. The way coal is processed depends upon its destination. Coal for steam production is handled somewhat differently than coal for metallurgical purposes.

A common practice is to wash the coal to remove impurities such as clay, rock, and slate. Since these will not burn, they are considered impurities. Many of these impurities are heavier than coal. In the washing process, coal can be floated and the impurities will sink to the bottom and can be removed. Washing increases the BTU value of the coal. One pound of unwashed coal can have as much as 2,000 BTU's less than when it is washed.

Before the coal is washed, it is usually crushed to further liberate any impurities which may have adhered to its surface. The coal is fed into a giant rotary or roll crusher where it is reduced to a size that will pass through a 6" round hole.

The coal is then "jigged." The crushed mineral is placed in enormous machines which process up to 1,200 short tons per hour. As this machine is cycled on and off, re-stratification occurs and the heaviest material ends up on the bottom. Layers of lighter material rest in strata above the heavier material. Basically there are three strata: refuse, clean coal, and "middlings." Because some of the coal contains more imbedded impurities it will neither sink to the bottom or float to the top. These "middlings" are crushed again and re-jigged.

The University of Kentucky S.E.C.C. Archives/International Harvester Collection

A coal preparation plant operator watches a piece of coal processing equipment at International Harvester's mines near Benham, Kentucky. 196?

Since all coal is different, sometimes many different processes must be employed to accomplish the cleaning process. One interesting invention involves "froth flotation." Fine particles of coal 1/2 mm and below are placed in a solution of water and chemicals which is churned so that it contains "froth" or air bubbles. The lighter particles attach themselves to the air bubbles and the heavier particles sink to the bottom.

Since the coal is wet when it finishes washing, it must be dried before it can be shipped. This is accomplished by several methods: fixed screens, vibrating screens, vast thickening tanks, vacuum filters, centrifugal dryers, and thermal dryers. Basically, these machines allow the coal to drain and

A steam engine pulls coal cars at a United States Steel mine near Lynch, Kentucky. 192?

the moisture to evaporate, the moisture to be spun out, or use heated air to complete the process.

Coal which has higher BTU and less impurities generally is blended with other coal to produce the right combination for metallurgical purposes. Coal with less BTU and higher impurities usually ends up being burned for electrical generation. Many coal companies employ chemists to determine the qualities of the coal they are mining. They determine the correct blend for the particular purposes for which the coal will be used.

After grading, washing and drying, the coal is loaded into rail gondolas and shipped to its destination.

Transportation

In the early days, coal was moved by carrying it in baskets. As a result, coal has sometimes been measured in "bushels." Buckets, tubs, boxes, and wheelbarrows were also used. Later, animals pulling wooden sleds were used to get the coal to the outside. Wooden plank roads were built to accommodate wheeled vehicles such as wagons. Wooden rails were employed to provide a bed for rail cars. Even the cars were wooden.

Metal soon became more popular because of its durability. Diesel engines, metal cars, and metal rails soon replaced the wooden ones. The advent of diesel and electric engines also brought a variety of rubber-tired vehicles used to transport coal. Today, mechanical conveyor belts augment the other transportation modes. A modest amount of coal moves from the mine mouth to the rail tipple by conveyor belt.

Moving the coal on the surface can be challenging, too. Coal pipeline "slurries" have been in use for some time and studies are underway to improve them so coal can be moved more easily from the west to the east. Some coal moves by motor carrier. These can be owner-operator trucks or independent trucking companies who are under contract.

Coal is unloaded into a tipple or preparation plant for grading and washing. The coal is then loaded into rail cars for long distance transport. More than 2/3 of the coal is moved either largely or entirely by rail.

Arch Coal, Inc.

Coal Truck

Towboat

Mining Lore

There are many good folk tales about mining. Miners and their families tell these stories again and again. Several of them involve the height of the coal. One man told me that he worked in a coal mine that was so low that he had to carry his lunch in a

Two miners pose with a Goodman electric motor at a United States Steel mine near Lynch, Kentucky (note the cloth hats and carbide lights).

fishing tackle box with sliding drawers so he could manage to open the box and get to his sandwiches. Another, not to be outdone, said that he worked in a seam that was so low he had to eat pancakes that had been dragged in with a piece of shot wire. My

daddy claimed that he worked in coal so low that he had to roll up to the face in order to shovel. If his timing was off, he would land with his back to the coal, and subsequently would have to keep trying until he landed with his shovel and face toward the fallen coal.

Some stories have to do with the "gob" rats who live in the mine and subsist on scraps from the miners' lunch buckets. Miners believe that the rats are good warning signals for rock falls. They claim that the rats can feel vibration in their sensitive feet and know when the top is about to collapse. They will leave the mine in droves just before a major rock fall.

My daddy claimed that he saw a "gob" rat wrap his tail around the lid of a round dinner pail while the other rat pulled him around in a circle until the lid was removed. The sandwiches were then stolen.

Miners can be practical jokers. My daddy told the story of a man who would wait until the miners got into the shower and lathered their heads and faces with soap before he threw a live snake in the shower room and yelled: "Snake!!!!! Snake!!!!"

As a method of friendly aggravation, miners often "cut" each other's lunch buckets. This means that if anyone finds your bucket lying unprotected, he may open it and eat what he wants.

Miners will sometimes use the epoxy from roofbolts to glue various objects. Lunch buckets and tools are favorite objects.

Sometimes older miners initiate younger miners. A cap board is used to dip grease from a five gallon bucket and the rookie is " greased" from head to toe. Sometimes electricity is hooked to his lunch bucket or shovel. The rookie is often asked to fetch

Melcroft Coal Company Camp Site at Coxton, Kentucky. This is an example of a corporate coal camp in the Appalachian mountains. May 16, 1923

items that do not exist, such as a "sky hook" or a "roof bolt for a strip mine."

Many times jokes are made about the rookie's intelligence—or lack of it. One underground miner, supposedly, shoved a cap light back across the lamp man's counter and said, "I won't be needing this, I'll be on the day shift." Another told a buddy who was driving a supply car to ". . . hurry past that bad place in the roof." He claimed that the company would be mad if he let a rock fall on the motor carrier.

Miners can be superstitious. For years, miners did not like the idea of women coming into a coal mine. They thought that women were bad luck and if they entered a mine, an explosion or rock fall would occur as a result. Some of the very first women coal miners were slaves who worked in the mines during the Civil War.

Back in the days of the oil lamps and carbide lights, miners used to think that their wives were going out on them if their light suddenly was extinguished.

There are many stories of ghost miners appearing underground, especially when mining occurs near old works. Miners sometimes see these apparitions in remote, isolated places in the mines.

The University of Kentucky S.E.C.C. Archives / International Harvester Collection

These men are hand excavating for an intermediate coal bin at an International Harvester mine near Benham, Kentucky. April 21, 1919.

Mining Songs

Archie Green, in his book *Only A Miner*, documents 36 songs which relate to coal mining. He and other folklorists suggest that American mining songs have played an important part in describing the plight of the coal miner to the world. They have served as warnings, political statements, and chronicles of the physical labor required to make the engine of these United States run smoothly. Most of these songs document mining from the laborer's view— the common, everyday worker who must travel into this dark, underground place filled with danger and work under extreme conditions to make a living. Many of them leave us with the feeling that coal miners were subjected to slave-like conditions, treated unfairly, and had little means to change their condition.

Many of these songs chronicle mine explosions, the addictive lure mining has for the miners, the idea that miners were treated as mules and not as humans, the toughness required to be a miner, and the terrible social and environmental destruction mining brings to the region.

The first coal mining song I can recall hearing was Western Kentucky native Merle Travis' song "Dark As a Dungeon." According to Archie Green, this song appeared in a recording session in the late 1940s along with "Nine Pound Hammer" and "Sixteen Tons" and was written under a street light near Redondo Beach while Travis was returning home from a date.

Starting a Coal Mine

Starting a coal mine costs a lot of money. Preparing a single seam for mining can cost approximately 10 million dollars. Because of the expense, the coal mine owner must carefully plan the entire operation. He must find out who owns the land. The land must either be purchased or leased. Many coal operators do not own the surface of the land. They purchase the mineral under the surface. The type of overburden (material lying above the coal) must be determined. They must calculate how close the coal is to the surface. All this can be calculated by using big "core" drills that have hollow bits. As the bit drills into the earth, the rock and coal that is fed into the center of the bit can be retrieved and analyzed. The quality of the coal can be determined from this sample by testing its BTU, coking qualities, sulfur content, and residual ash.

The mine owner must decide what method he will use to extract the coal, how he will transport it, and who will buy his product. The methods used to extract coal depend on the depth of the coal bed, size (height), flow of the coal seam, and structural conditions of the type of earth between the seam and the surface terrain. Coal usually is transported long distances by rail or truck. Surveying, mapping, and road construction must occur before the actual coal is accessed and mined. Millions of dollars of equipment must be purchased. Experienced miners must be found to fill supervisory positions. Trained miners with a variety of skills must be located and hired. Tipples, rail, and support buildings must be built. Government permits must be applied for and approved. A plan for the mining and reclamation must be filed with the governing agencies. Money has to be posted as a bond to insure the restoration of the site.

The University of Kentucky S.E.C.C. Archives / International Harvester Collection

This man is checking the switch panel for an electric generator at International Harvester's mine near Benham, Kentucky. 193?

The Future of Mining

The future of coal is very positive. Our nation imports over 60% of its oil but almost none of its coal. Most experts contend that coal's place in the world market is for residential and commercial production of electricity and for metallurgical processes. Although the technology has been developed for coal liquefaction and gasification, current oil prices are so low that coal cannot compete with oil for uses in transportation.

Mining equipment is getting better and better. Machines are more reliable and mining companies have changed their attitudes toward maintenance. Much like the aviation industry, new maintenance schedules now provide for the machinery to be replaced at preplanned intervals. Mining companies have realized that if they wait until the machine breaks down the cost can approach $75,000 per hour. These businesses can no longer afford for their machinery to be out of production.

New technology in longwall mining has allowed the industry to maintain production levels, even with a reduction in the number of active machines. In 1989, there were 112 longwall mining machines operating in the United States. Currently, there are significantly higher production levels with only 79 machines in operation.

According to Dr. Kot Unrug, Professor of Mining Engineering at the University of Kentucky, current research is in the area of thin seam mining. There are approximately another 150 years of mining remaining in these smaller seams, if the technology develops and is economical. Plans are to develop machinery which will use chain/plow

Coal washer at Totz, Kentucky, 1995.

James L. Goode, Photographer

Excavator, truck, and endloader at Pine Branch Coal Company, April 29, 1995

technology to mine the coal by remote control using video cameras.

Two problems must be solved: 1) seismic and radio tomography must consistently report the geology of the potential seam; 2) the speed at which the machine operates must be increased.

There will be advantages to this new system: 1) There will be no personnel in the tunnel where mining is occurring; 2) Machinery will be maintained through service tunnels which will be dust free; 3) Most all available coal resources will be utilized to meet future energy needs.

Dr. Unrug and his staff are developing computer programs which will allow improvements in safety and enhance responses to emergency situations. Computer simulations will present various scenarios related to environmental conditions, air flow, smoke

patterns, temperature fluctuations, and the presence or absence of various gases. These simulations will provide engineers with a set of probability factors which can be used to prevent or respond to mining accidents. Mining companies are being encouraged to place their physical plans and models of their mine on computers which can then be used in concert with the University computer simulations to provide officials with critical information.

New roof control systems are being used by a few mining companies in Utah, Wyoming, and West Virginia. A cable bolt system, which is similar to solid steel glue bolts but uses flexible steel cable instead, is being used in longwall sections in a few mines. These "bolts" are easier to feed into the drill holes when the overburden requires longer bolts but the seam is too low to allow rigid bolts to be placed without bending and re-bending them. These new bolts have also been helpful in building other kinds of roof control devices such as trusses.

In a few German mines a new longwall plow is being used. The Heitzman System involves a series of chains placed 1/2 meter apart and equipped with specially designed metal points which are pulled along to plow the coal from the face. Another chain structure carries the coal to the gate entry where it is transported to the surface.

The future of mining will continue to improve as more "real time" computer applications are made, developing and developed technology makes low seam mining more economical, more sophisticated maintenance programs are used, stability and ventilation problems are solved, and the demand for electricity and metallurgical products increases.

Mining Terms

Agglomerating ~ The ability of coal to "cake" or "coke" when heated

Air Course ~ The tunnel through which air is circulated

Anemometer ~ A small, hand-held machine used to measure the amount of air in cubic-feet-per-minute

Barrier Pillar ~ A solid block of coal left between entries as support for weight from the mine roof—sometimes left between two mines or sections within a mine

Black Damp ~ A mixture of carbon dioxide and oxygen

Brattice ~ Any partition made of cloth, wood, or stone used to direct air flow to the working place

Breaker Boys ~ Young boys, often 9-10 years old, who broke up larger blocks of coal or picked rock and slate from the coal as it exited the mine.

Breakthrough ~ An opening cut through a pillar of coal into another room

Burden ~ In strip mining, the distance between rows of holes which have been drilled for blasting overburden

Cap ~ A device for igniting explosives

Cap Board ~ A wedged shaped piece of wood which is used to drive into spaces above timbers to tighten them. Also used as a general method of wedging two objects.

Carboniferous Period ~ A geologic division of the Paleozoic era which includes the Mississippian and Pennsylvanian. During the Pennsylvanian period, swamps were formed and plant remains were deposited which were eventually changed into coal.

Collier ~ A ship which has coal as its cargo

Continuous Miner ~ A method of mining coal which involves tearing or ripping the coal from the face by mechanical means and loading it onto shuttle cars. This eliminates the conventional mining practice of undercutting, drilling, shooting, and loading.

Conventional Roof bolt ~ A metal bolt of varying lengths used to secure the roof of the mine. This type is threaded on the end and uses a "shuck" or type of nut that expands as it is screwed onto the bolt.

Crib ~ This is a method of roof support involving 6" square short lengths of wood arranged in a log cabin pattern.

Cross Collar ~ a square length of wood placed horizontally against the roof with round timbers supporting each end to add strength to the roof, especially over roadways.

Curtain ~ A sheet of plastic, canvas, or burlap (brattice cloth) used to direct the flow of air in a mine

Dinner Hole ~ A place where miners gather to eat lunch. This location is usually near a light and often adjacent to a power center which generates heat from its transformers.

Modern miners often have microwave ovens at these "dinner holes" to heat their lunches.

Draw Rock (Slate) ~ The soft slate or rock above a coal seam which falls easily when the coal is removed.

Drift Mine ~ Usually the type mine that is begun by entering the coal seam horizontally and advancing through the seam along its contours.

Drill Chips ~ In strip-mining, the bench drills produce dust or particles while penetrating the rock commonly called "stemmings." These are used to tamp the last few inches of the hole, after it is loaded with the cap, primer cord, and diesel fuel/ammonium nitrate.

Face ~ The wall of coal where the underground mining production activity is taking place

Flame Safety Lamp ~ A specially designed lamp used to check oxygen levels and the presence of methane or carbon dioxide

Firedamp ~ The combustible gas methane. The combustibility factor ranges from pure methane to mixtures with air at 5 to 15 percent levels.

Gasification ~ Combustible gas produced by burning coal underground

Glue Bolt or Resin Bolt ~ A type of roof bolt which utilizes the adhesive quality of resin to prevent rock falling from the roof

Gob Rats ~ Rats that live underground in the "Gob" or sections along the rib where most of the trash collects

Gondolas ~ Various sizes of rail cars used to transport coal

Longwall Mining ~ In this practice, the coal is cut into a large block with continuous mining machines and a larger machine, similar to a meat slicer, then slices or "shears" the coal from the short side using a drum equipped with carbide-tipped bits. This machine "walks" on hydraulic legs and does not require roof bolting in the newly mined area.

Liquefaction ~ Coal which has been altered, by chemical and mechanical processes, into a liquid

Man trip ~ A rubber tire vehicle or a rail car used to transport men into the working portion of the active mine

M.S.H.A. ~ U.S. Mine Safety and Health Administration founded in 1973

Methane Monitor ~ Devices which monitor the levels of methane in the circulating air within a coal mine

Mine Run Coal ~ Raw coal as it exits the mine. This is dirty coal, before washing and processing has removed rock, slate, dirt, and other impurities.

O.S.H.A. ~ U. S. Occupational Safety and Health Administration

O.S.M. ~ U. S. Office of Surface Mining

Overburden ~ The layer of rock, slate and dirt covering the coal seam

Peat ~ Decayed organic matter which has been buried in the absence of oxygen—the earliest stage of organic matter which eventually is transformed, through a chemical reaction, into coal if subjected to aging, pressure, and the absence of oxygen

Photosynthesis ~ A process by which chlorophyll-containing cells in green plants convert sunlight to chemical energy

Primer Cord ~ Hollow plastic cord which is filled with fine black powder and used in strip mining for non-electrical detonation of explosive charges

Ramping Down ~ In strip-mining, this is the process of removing overburden in a manner so as to create a roadway which slopes into the coal bed.

Ratio Feeder ~ A device which breaks up the coal into more manageable pieces and discharges it onto a conveyor belt at a fixed rate so as to not overload it

Reclaimed Land ~ Land that has been mined and has been restored to original contour and had substantial vegetation established

Return Air ~ The air that is returned or exhausted to the exterior of the mine after it has been circulated across the face where miners are working

Rib ~ The walls of solid coal along any passage in a coal mine

Rock Dust ~ Limestone dust which

is applied dry or wet on the roof, bottom, rib, and face to render coal dust inert and prevent explosions

Room & Pillar ~ A method of underground mining which creates "rooms" and leaves "walls" or pillars for roof support

Scoop ~ A rubber tire vehicle which is usually powered by a D. C. motor and has a small wedge-shaped bucket on the front for cleaning the bottom and hauling supplies

Section Crew ~ A group of miners who work together in one particular area of the mine. This crew basically includes a face boss, continuous miner operator, continuous miner helper, shuttle car drivers, roof bolters, belt men, brattice men, supply men, and general inside laborers.

Self-Rescuer ~ A device carried by the miner on his work belt, which, when deployed, chemically filters carbon monoxide present in the air so that he is able to inhale safely

Shaft Mine ~ An entry to the coal gained by sinking a vertical shaft down to the mineable seam and then advancing horizontally

Shear ~ The part of a longwall mining machine which consists of a large, cylindrical drum equipped with carbide tipped bits. This is used to "rip" or "shear" the coal from the working face.

Shuttle Car ~ A rubber-tired vehicle used to transport coal from the continuous miner or loading machine to a ratio feeder which then distributes it to a conveyor belt

This photograph depicts a mock accident scene in a coal mine at Benham, Kentucky.

Slope Mine ~ An entry to the coal gained by advancing a tunnel at a sloping 10 to 15 degree angle through overburden

Spacing ~ In strip-mining, the distance between holes drilled for explosive charges

Stemmings ~ Dust and particles which exit a drill hole as the bit advances

Stink Damp ~ Hydrogen Sulfide gas which sometimes is released by coal

Stopping ~ Any air-tight wall built across a tunnel in a mine which is used to direct air flow, redirect the flow of water, or create a barrier between active parts of the mine and sections where mining has been completed

Stump ~ The remainder of a barrier pillar which has been "robbed" or mined as much as the roof will allow

Tipple ~ The holding bin where coal is received from underground and processed for shipping

Unit Train ~ A collection of connected rail cars which transports coal to a common destination. These often consist of one hundred, 200-ton rail cars pulled by several diesel locomotives.

White Damp ~ Carbon monoxide gas which can emerge in coal mines underground

A coal miner who lost his arm in an accident plays a guitar with a makeshift noter.

Bibliography

Adler, Irving and Ruth. *Coal.* New York: The John Day Company, Inc., 1965.

American Mining Congress. Mining: Discoveries For Progress *Tell Everyone About Mining* (Video). Washington, D.C.: American Mining Congress, 1993.

Arnold, Guy. *Coal.* New York: Gloucester Press, 1985.

Bailey, Rufus and Katrine. *Black Diamonds In My Own Back Yard.* Collegedale, Tennessee: The College Press, 1979.

Berkowitz, Norbert. *The Chemistry Of Coal.* New York: Elsevier, 1985.

Burt, Olive. *Black Sunshine: The Story of Coal.* New York: Julian Messner, 1977.

Campbell, Shirley Young. *Coal & People.* Parsons, West Virginia: McClain Printing Company, 1994.

Caterpillar Corporation. *Common Ground* (Video). Nashville: Capital Communications, 1993.

Chaffin, Lillie D. *Coal Energy And Crisis.* New York: Harvey House, 1974.

Chakravarthy, Balajis. *Managing Coal: A Challenge In Adaptation.* Albany, N.Y.: State University of New York, 1981.

Davey, John. *Coal Mining.* London: A&C Black, Ltd, 1960.

Davis, Bertha. *The Coal Question.* New York: F. Watts, 1982.

Dionetti, Michelle. *Coal Mine Peaches.* New York: Orchard Books, 1991.

Doty, Roy. *Where Are You Going With That Coal?* New York: Doubleday & Co., Inc., 1977.

Francis, Wilfred. *Coal: Its Formation And Composition.* London: E. Arnold, 1961.

Goode, James B. *Coal, Steel, Machines, & Men. The Benham Story.* Benham, Kentucky: The University of Kentucky Southeast Community College, 1993.

Goode, James B. *Lynch: A Coal Legacy.* Benham, Kentucky: The Kentucky Humanities Council, 1990.

Goode, James B. *Up From The Mines.* Ashland, Kentucky: The Jesse Stuart Foundation, 1993.

Green, Archie. *Only A Miner: Studies in Recorded Coal Mining Songs.* Urbana: University of Illinois Press, 1972.

Harter, Walter L. *Coal: The Rock That Burns.* New York: E.P. Dutton, 1979.

Harvey, Curtis E. *Coal In Appalachia.* Lexington, Kentucky: The University Press of Kentucky, 1986.

Hower, Judith M. *Coal In Kentucky: An Overview.* Lexington, Kentucky: Kentucky Department of Energy, 1981.

I. E. A. Coal Research. *Coal Thesaurus 1978.* London: I.E.A. Research, 1978.

Kamecke, Theo. *Coal: Bridge To The Future.* Woodbury, N.Y.: J.N. Company, 1980.

Korson, George Gershon, Editor. *Coal Dust On The Fiddle: Songs And Stories of the Bituminous Industry.* Philadelphia: University of Pennsylvania Press, 1943.

Lindberg, Kristina and Barry Provorse. *Coal: A Contemporary Energy Story.* Seattle: Scribe Publishing Corporation, 1977.

Madisonville Community College. *Coal Miner's Jargon.* Big Stone Gap, Virginia: Madisonville Community College, 19?.

Miller, Donald L. & Richard E. Sharpless. *The Kingdom of Coal: Work, Enterprise,& Ethnic Communities in the Mine Fields.* Philadelphia: University of Pennsylvania Press, 1985.

Mitgutsch, Ali. *From Swamp to Coal.* Minneapolis: Carolrhoda Books, 1985.

Newell, John. *Behind The Scenes: In A Coal Mine.* Great Britain: J. M. Dent & Sons, Ltd., 1964.

North Cambria High School. *Out of the Dark 2: Mining Folk.* Indiana, Pennsylvania: The A.G. Halldin Publishing Company, 1977.

Rickert, D.A., W. J. Ulman, and E. R. Hampton, eds. *Synthetic Fuels Development: Earth Science Considerations.* Washington, D.C.: U.S. Department of the Interior/ Geological Survey, 1979.

Ridpath, Ian. *Man & Materials: Coal.* Reading: Addison-Wesley, 1975.

Ritchie, Jean. *Black Waters. Clear Waters Remembered.* Sire SES 97014, 1970.

Rogers, William Patrick. *The Coal Primer: A Reference Handbook For Non-Coal People.* Little Rock, Arkansas: William P. Rogers, 1978.

Rowlands, Dorothy Howard. *Coal And All About It: The Romance Of A Great Industry.* London: Harrap, 1946.

Schmidt, Richard A. *Coal In America: An Encyclopedia of Reserves, Production, and Use.* New York: Coal Week, McGraw-Hill Publications, Co., 1979.

Stewart, Gail. *Coal Miners: At Risk.* Mankato, Minnesota: Crestwood House, 1988.

Theiss, Nancy. *King Coal In Kentucky.* Louisville, Kentucky: The Courier Journal & Louisville Times, 1985.

Tomalin, Miles. *Coal Mines and Miners.* New York: Roy Publishers, 1960.

United Mine Workers of America Journal: Special Bicentennial Issue. 11: 15 June 1976.

United States Bureau of Mines. *Coal Country* (Computer Program). Washington, D. C.: Office of Public Information/ U.S. Department of the Interior/ U. S. Bureau of Mines, 1994.

United States Department of Labor/ Mine Safety and Health Administration/ National Mine Health and Safety Academy. *Coal Mine Maps: Safety Manual No. 12.* Washington, D.C.: U.S. Government Printing Office, 1994.

United States Department of Labor/ Mine Safety and Health Administration/ National Mine Health and Safety Academy. *Coal Mine Roof and Rib Control: Safety Manual No. 18.* Washington, D.C.: U.S. Government Printing Office, 1994.

United States Department of Labor/ Mine Safety and Health Administration/ National Mine Health and Safety Academy. *Coal Mining: Safety Manual No. 1.* Washington, D.C.: U.S. Government Printing Office, 1987.

United States Department of Labor/ Mine Safety and Health Adminis-

tration/ National Mine Health and Safety Academy. *Introduction to Underground Coal Mining.* Washington, D.C.: U.S. Government Printing Office, 1986.

United States Department of Labor/ Mine Safety and Health Administration/ National Mine Health and Safety Academy. Mine Escapeways: *Safety Manual No. 11.* Washington, D.C.: U.S. Government Printing Office, 1994.

United States Department of Labor/ Mine Safety and Health Administration/ National Mine Health and Safety Academy. *Mine Gases: Safety Manual No. 2.* Washington,D.C.: U.S. Government Printing Office, 1994.

United States Department of Labor/ Mine Safety and Health Administration/ National Mine Health and Safety Academy. *Mine Ventilation: Safety Manual No. 20.* Washington, D.C.: U.S. Government Printing Office, 1994.

United States Department of Labor/ Mine Safety and Health Administration. *M.S.H.A.Commemorates 25th Anniversary of Coal Act* (Filename: MSHA95.110). Washington, D.C.: Mine Safety and Health Administration, 1995.

United States Department of Labor/ Mine Safety and Health Administration. *Historical Data on Mine Disasters in the U. S.* (Fact Sheet No. M.S.H.A. 93-9). Washington, D.C.: Mine Safety and Health Administration, 1992.

United States Department of Labor/ Mine Safety and Health Administration. *Injury Trends in Mining* (M.S.H.A. Fact Sheet No. M.S.H.A. 93-2). Washington, D.C.:Mine Safety and Health Administration, 1992.

United States President's Commission On Coal. *Coal Data Book: The*

President's Commission On Coal. Washington, D.C.: The U.S. Government Printing Office, 1980.

Ward, Colin R. *Coal Geology and Coal Technology.* Boston: Blackwell Scientific, 1984.

Wolfgang, Paul. *Mining Lore.* Portland, Oregon: Morris Printing Company, 1970.

Worker's of the Writer's Program of the Work Projects Administration in the Commonwealth of Pennsylvania. *The Story of Coal.* Chicago: Albert Whitman & Co., 1944.

Zabetakis, Michael G., and Joseph J. Yancik. *Coal.* Monroeville, Pennsylvania: Bituminous Coal Research, Inc., 1981.

Zieger, Robert H. *John L. Lewis.* Boston: Twayne Publishers, 1988.

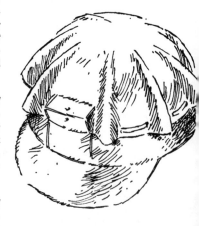

Early Miner's Helmet